SMF/AMS TEXTS *and* MONOGRAPHS

Panoramas et Synthèses • *Numéro 2* • 1996

Mirror Symmetry

Claire Voisin

Translated by
Roger Cooke

American Mathematical Society
Société Mathématique de France

1991 *Mathematics Subject Classification.* Primary 14D05, 14D07;
Secondary 14J32, 14M25, 32G13, 32G20, 32L07, 81T30, 81T40, 53C23, 53C15.

ABSTRACT. This book describes recent works motivated by the discovery of the mirror symmetry phenomenon by the physicists. The first chapter is devoted to the geometry of Calabi–Yau manifolds, and the second describes, as motivation, the ideas from quantum field theory that led to this discovery.

The other chapters deal with more specialized aspects of the subject: the work of Candelas, de la Ossa, Greene, and Parkes, based on the fact that under the mirror symmetry hypothesis the variation of Hodge structure of a Calabi–Yau threefold determines the Gromov–Witten invariants of its mirror; Batyrev's construction, which exhibits the mirror symmetry phenomenon between hypersurfaces of toric Fano varieties, after a combinatorial classification of the latter; the mathematical construction of the Gromov–Witten potential, and the proof of its crucial property (that it satisfies the WDVV equation), which makes it possible to construct a flat connection underlying a variation of Hodge structure in the Calabi–Yau case. The book concludes with the first "naive" Givental computation, which is a mysterious mathematical justification of the computation of Candelas et al.

Library of Congress Cataloging-in-Publication Data
Voisin, C. (Claire)
  [Symétrie Miroir. English]
  Mirror symmetry / Claire Voisin ; translated by Roger Cooke.
    p. cm. — (SMF/AMS texts and monographs, ISSN 1525-2302 ; v. 1) (Panoramas et synthèses ; Numéro 2, 1996)
  Includes bibliographical references.
  ISBN 0-8218-1947-X
  1. Mirror symmetry. 2. Quantum field theory. 3. Manifolds (Mathematics) I. Title. II. Series. III. Series: Panoramas et synthèses ; 2.
  QC174.17.S9V6213 1999
  530.14′3—dc21   99-35675
                                                                                    CIP

Copying and reprinting. Individual readers of this publication, and nonprofit libraries acting for them, are permitted to make fair use of the material, such as to copy a chapter for use in teaching or research. Permission is granted to quote brief passages from this publication in reviews, provided the customary acknowledgment of the source is given.

Republication, systematic copying, or multiple reproduction of any material in this publication is permitted only under license from the American Mathematical Society. Requests for such permission should be addressed to the Assistant to the Publisher, American Mathematical Society, P. O. Box 6248, Providence, Rhode Island 02940-6248. Requests can also be made by e-mail to reprint-permission@ams.org.

© 1999 by the American Mathematical Society. All rights reserved.
The American Mathematical Society retains all rights
except those granted to the United States Government.
Printed in the United States of America.

∞ The paper used in this book is acid-free and falls within the guidelines
established to ensure permanence and durability.
Visit the AMS home page at URL: http://www.ams.org/

10 9 8 7 6 5 4 3 2 1   04 03 02 01 00 99

Mirror Symmetry

Contents

Introduction	vii
Organization of the text	xviii
Acknowledgment	xviii
Note added in Translation	xix
Chapter 1. Calabi–Yau Manifolds	1
1. Yau's Theorem	1
2. The decomposition theorem	3
3. Smoothness of the local family of deformations	5
4. Smoothability of Calabi–Yau manifolds with normal crossings	8
5. The period map	10
6. Calabi–Yau threefolds	14
7. Examples of Calabi–Yau manifolds	15
8. Mirrors	17
Chapter 2. "Physical" origin of the conjecture	21
1. The $N=2$-supersymmetric σ-model	21
2. Quantification	27
3. Gepner's conjecture	31
4. Mirror symmetry	32
5. The $N=2$-superconformal theory and Dolbeault cohomology	33
6. Witten's interpretation	35
Chapter 3. The Work of Candelas–de la Ossa–Green–Parkes	39
1. Special coordinates and Yukawa couplings	39
2. Degenerations	43
3. The Candelas–de la Ossa–Green–Parkes calculation	48
4. Picard–Fuchs equations	50
5. Conclusion of the argument	54
Chapter 4. The work of Batyrev	57
1. Toric varieties	57
2. Weil and Cartier divisors	59
3. Polyhedra and toric varieties	60
4. Toric Fano varieties	62
5. Desingularization	63
6. Calculation of the cohomology of $\widehat{Z_f}$	65
Chapter 5. Quantum cohomology	73
1. The formulation by Kontsevich and Manin	73

- 2. The work of Ruan and Tian — 76
- 3. Gromov–Witten potential — 81
- 4. Application to mirror symmetry — 87
- 5. Quantum product — 88
- 6. The calculation of Aspinwall and Morrison — 89

Chapter 6. The Givental Construction — 95
- 1. Floer Cohomology — 95
- 2. The comparison theorem — 101
- 3. Quantum cohomology and Floer cohomology — 102
- 4. Equivariant cohomology — 105
- 5. The Givental construction — 110

Bibliography — 117

Introduction

The present book consists of a set of notes from a course given by me at the Institut Henri Poincaré in the context of the algebraic geometry semester at the Émile Borel Center during the spring of 1995.

The goal of the course was to present recent results connected with mirror symmetry. As this topic is far from being perfectly understood mathematically, these recent results have developed in a number of different directions, from the profound study conducted by Batyrev on families of hypersurfaces with trivial canonical bundle on toric Fano varieties and the combinatorial construction of the mirror family to the discovery of the "quantum product" in the cohomology of a symplectic manifold. This last topic goes far beyond the scope of mirror symmetry, but it was motivated by the desire to give a mathematical definition of such objects as the Gromov–Witten potential, which is one of its essential ingredients, and to prove the main property of that potential, namely that it satisfies the "WDVV" equation.

This burgeoning situation is reflected in the division of the book into autonomous chapters, which, although they are connected by common themes (variation of Hodge structure, Calabi–Yau manifolds and their rational curves, and naturally, mirror symmetry) do not necessarily involve any very intimate logical connection.

In this introduction I propose nevertheless to give a synthesis of the subject intended to orient the book around the subject of mirror symmetry and also to show that the mathematical papers to which it has given rise, which are described in this book, interesting though they are intrinsically, are far from providing a justification as tangible as that proposed by physicists in the language of field theory (and unfortunately on the basis of quantum formalism and Feynman integrals, which seem incapable of being rigorously justified).

A *Calabi–Yau manifold* is a compact complex Kähler manifold X having trivial canonical bundle, that is, possessing a holomorphic form η that never vanishes, belonging to $H^0(X, \overset{n}{\wedge} \Omega_X^n)$, where Ω_X is the holomorphic cotangent bundle of X and where $n = \dim X$.

Throughout the following we assume that

$$H^2(\mathcal{O}_X) = \{0\},$$

even though some interesting studies have been conducted in the case of K3 surfaces (that is, simply connected Calabi–Yau twofolds), for which this assumption does not hold. Under this condition the Kähler cone of X is open in $H^2(X, \mathbb{R})$, and it is possible to introduce the "complexified Kähler cone" of this manifold, which is one of the aspects of the moduli space used by physicists. This cone is the open set

$$K(X) \subset H^2(X, \mathbb{C})/2i\pi H^2(X, \mathbb{Z})$$

defined by the condition

$$\omega \in K(X) \iff \operatorname{Re}\omega \text{ is a Kähler class.}$$

Mirror symmetry consists essentially of the existence of a mirror family $\{X'\}$ of Calabi–Yau manifolds of the same dimension such that $K(X)$ uniformizes (that is, is a covering of) the moduli space Def X of deformations of the complex structure of X' and $K(X')$ uniformizes that of X. The exact nature of this covering is not completely understood, but it should be provided by a partial "marking" of the cohomology of X' (resp. X).

The elliptic curves ($n = 1$) provide the simplest example of this phenomenon: the complexified Kähler cone $K(E)$ (which is independent of the complex structure of E) can be canonically identified, through integration over E, with the set

$$\{\lambda \in \mathbb{C}/2i\pi\mathbb{Z};\ \operatorname{Re}\lambda > 0\}.$$

On the other hand, the set of marked complex structures on E can be identified by the period map with the set

$$\mathcal{H} = \{\tau \in \mathbb{C}|\ \operatorname{Im}\tau > 0\}$$

and the translations by $u \in \mathbb{C}$ on this set correspond simply to the action of upper-triangular integral matrices with diagonal equal to the identity on the set of markings of an elliptic curve E (that is to say, in this case, the symplectic isomorphisms $H^1(E,\mathbb{Z}) \cong \mathbb{Z}^2$ endowed with the standard symplectic form). The map

$$K(E) \longrightarrow \mathcal{H}/\mathbb{Z},$$

$$\lambda \mapsto \tau = -\frac{\lambda}{2i\pi} \operatorname{Mod}\mathbb{Z}$$

thus gives a uniformization of the kind predicted by mirror symmetry.

In higher dimensions a new phenomenon appears: the family $\{X'\}$ is generally different from the family $\{X\}$; for the underlying topological manifolds are different simply because their Betti numbers are different. However, the Hodge numbers of X', that is, the numbers

$$h^{p,q}(X') := \dim H^{p,q}(X') := \dim H^q(X', \Omega_{X'}^p),$$

can be derived from the Hodge numbers of X as follows.

At a point $\omega \in K(X)$ having image X' the uniformization

$$K(X) \longrightarrow \text{ moduli space of } X'$$

induces an isomorphism

$$H^1(\Omega_X) \cong H^1(T_{X'})$$

on the level of the tangent spaces, where we have used the natural identifications

$$T_{K(X),\omega} \cong H^2(X, \mathbb{C}) \cong H^1(\Omega_X),$$

the second of which is a consequence of the assumption $H^2(\mathcal{O}_X) = \{0\}$.

More generally, as follows from the "construction" of mirror symmetry by physicists, for every p and q we should have isomorphisms

$$H^q(\Omega_X^p) \cong H^q(\overset{p}{\wedge} T_{X'}).$$

Finally, the choice of a holomorphic n-form $\eta \in H^0(K_X)$ (where the form η is unique up to a constant multiple) determines isomorphisms given by the inner product
$$H^q(\overset{p}{\wedge} T_{X'}) \cong H^q(\Omega_{X'}^{n-p}),$$
which, when composed with the preceding isomorphisms, provide non-canonical isomorphisms
$$H^q(\Omega_X^p) \cong H^q(\Omega_{X'}^{n-p})$$
and hence a series of equalities:
$$h^{p,q}(X) = h^{n-p,q}(X').$$

We recall finally that according to Hodge theory the following direct-sum decomposition holds for each k:
$$H^k(X) = \bigoplus_{p+q=k} H^{p,q}(X).$$
This decomposition yields the relation between the Hodge numbers and the Betti numbers of X:
$$b_k(X) = \sum_{p+q=k} h^{p,q}(X).$$

When $n = 3$, comparison of the Hodge numbers of X and X' yields a comparison of the Betti numbers. Indeed:
- the assumption $H^2(\mathcal{O}) = \{0\}$ implies $b_2 = h^{1,1}$;
- on the other hand we have $h^{3,0} = 1$ and $h^{2,1} = h^{1,2}$ (since $H^1(\Omega_X^2)$ and $H^2(\Omega_X)$ are duals of each other); hence $b_3 = 2 + 2h^{2,1}$.

Finally, the relation $H^2(\mathcal{O}) = \{0\}$ is equivalent by Serre duality to the relation
$$H^1(K_X) = H^1(\mathcal{O}_X) = \{0\}$$
and hence to the relation $b_1(X) = 0$, so that we have
$$\begin{cases} b_1(X') = 0, \\ b_2(X) = \frac{1}{2}(b_3(X) - 2) = b_4(X'), \\ b_3(X') = 2 + 2b_2(X). \end{cases}$$

The constructions of mirror families available at present reduce essentially to projective algebraic geometry, the most general one being that of Batyrev, which involves partial desingularizations of hypersurfaces with trivial canonical bundle on toric Fano varieties. The latter are compactifications of $(\mathbb{C}^*)^{n+1}$ with ample anticanonical bundle to which the natural action of $(\mathbb{C}^*)^{n+1}$ on itself can be extended.

Batyrev has shown that these manifolds are in one-to-one correspondence with convex polyhedra with integer vertices in $\mathbb{R}^{n+1}$ having 0 as their only interior lattice point, and such that the dual polyhedron also has integer vertices (the so-called *reflexive property*). The mirror family is then obtained by partial desingularization—unfortunately the desingularization procedure is generally not unique—of the family of hypersurfaces with trivial canonical bundle on the toric Fano variety associated with the dual polyhedron.

The best-known example of such a construction is that of the physicists Candelas, de la Ossa, Green, and Parkes in the outstanding paper [43], which is admirably expounded in the language of mathematicians by Morrison [53]. One considers the

family of Calabi–Yau threefolds given by smooth hypersurfaces of degree 5 in $\mathbb{P}^4$; such a hypersurface has the following Hodge numbers:
$$h^{1,1} = 1, \quad h^{2,1} = 101.$$
The mirror family must therefore have the Hodge numbers
$$h^{1,1} = 101, \quad h^{2,1} = 1,$$
and the number of parameters for the deformations of the complex structure on this family must be 1.

This family is the following: consider the quintic polynomials (depending on a complex parameter $\lambda \in \mathbb{C}$) of the form
$$F_\lambda = \sum_{i=0}^{4} X_i^5 + \lambda X_0 \cdots X_4.$$
Each polynomial F_λ is invariant under the group
$$G = (\mathbb{Z}/5\mathbb{Z})^5/\mathrm{diag},$$
which acts on $\mathbb{P}^4$ by multiplying coordinates by a fifth root of unity.

The subgroup $H \subset G$ defined by the condition
$$(\alpha_0, \ldots, \alpha_4) \in H \Leftrightarrow \sum_i \alpha_i = 0 \quad \text{in} \quad \mathbb{Z}/5\mathbb{Z}$$
acts on $X_\lambda := \operatorname{div} F_\lambda$, and the action induced on $H^{3,0}(X_\lambda)$ is trivial.

It can thus be shown that the quotient X_λ/H admits a natural desingularization which is a Calabi–Yau desingularization. The family $\{\widetilde{X_\lambda/H}\}$ of dimension 1 is the required mirror family.

Morrison has expounded the calculus introduced in [**43**] from the point of view of Hodge theory; the essential ingredients are the following. A canonical coordinate at infinity is sought on a curve with coordinate λ (actually λ^5). In general such natural coordinates exist on the moduli space of a Calabi–Yau manifold (or rather the Kuranishi family, which is smooth) and depend on the choice of a partial marking of the cohomology. The point is that the monodromy around infinity provides such a marking naturally.

The logarithm t of the coordinate q thereby produced is then assumed to coincide through the mirror map with the natural coordinate existing on the complexified Kähler cone of the original family. The second point is that the same partial marking of the cohomology makes it possible likewise to trivialize the bundle $\mathcal{H}^{3,0}$ of rank 1, whose fiber at the point λ is the vector space $H^{3,0}(\widetilde{X_\lambda/H})$. Thus we have a function of q given by the value of the "Yukawa couplings" (that is, a cubic form on the tangent space to the family), normalized by the trivializing section of the bundle $\mathcal{H}^{3,0}$ over the field of logarithmic vectors $q\partial/\partial q$. The power series expansion is explicitly calculable and can be derived immediately from that of certain solutions of the Picard–Fuchs equation from the family $\{X_\lambda\}$.

The extraordinary mathematical novelty of this article thus lies in the identification of this series $\psi(q)$, where $q = e^t$, with the series
$$5 + \sum_{d>0} N(d) \frac{e^{td}}{1 - e^{td}}$$

where $N(d)$ is the number of immersed rational curves of degree d on a general quintic of $\mathbb{P}^4$. (This number should be finite according to a conjecture of Clemens.)

At the time these notes were written the prediction thereby obtained for the $N(d)$ had been verified for $d \leq 4$, which is already quite remarkable given the astronomical allure of these numbers. We note on the other hand that, despite the progress made by Kontsevich [**51**] on the problem of evaluating the $N(d)$, there were not at the time any methods making it possible to calculate these numbers except one by one.

A major achievement in the story of mirror symmetry is the work [**G**], where Givental proves, among other things, the correctness of these predictions.

The identification of these two series is a consequence of the "physical" construction of mirror symmetry. Assigning a complex structure on X and a complexified Kähler parameter $\omega = \alpha + i\beta$ determines via Yau's theorem a Kähler–Einstein metric of Kähler class α, while to β there corresponds a class of 2-forms that are closed modulo exact forms and forms that are an integral multiple of $2i\pi$ on every cycle of X of dimension 2. These assignments make it possible to construct an "$N = 2$-supersymmetric" σ-model given by an action $S(\phi, \psi)$ where ϕ is a mapping of a Riemann surface Σ onto X, and ψ is a section of $\phi^*(T_x) \otimes S$, S being the spinor bundle corresponding to the choice of a Spin structure on Σ.

The dependence of this action relative to the choice of a form $\tilde{\beta}$ representing β hardly affects the theory, since $\tilde{\beta}$ contributes to S only in the term $\int_\Sigma \phi^*(\tilde{\beta})$, so that a different choice $\tilde{\beta}'$ modifies the action by a "boundary term" and a term that assumes as values multiples of $2i\pi$ on surfaces without boundary. The boundary terms do not contribute to the Euler–Lagrange equations describing the critical points of S, and on the other hand only the exponential of the action appears, for example, in the calculation of correlation functions.

The action $S(\phi, \psi)$ is invariant (modulo a boundary term) by an infinite-dimensional Lie superalgebra of infinitesimal transformations, the "Virasoro $N = 2$-superalgebra," whose odd part, made up of "supersymmetric transformations" is better understood if one represents $S(\phi, \psi)$ as an expansion in components of an action $S(\Phi)$ associated with superdifferentiable maps Φ of a Riemann super-surface Σ on X. The even part of this Lie superalgebra is made up of two copies of the Virasoro algebra and reflects the conformal invariance of the action $S(\phi, \psi)$. The $N = 2$-supersymmetry of this action is a consequence of the fact that the metric is a Kähler metric.

The physicists wish to represent by quantification certain functionals on the space of classical solutions of the Euler equations, called "observables", on a Hilbert space. The construction of such a representation would be equivalent to assigning the "correlation functions" of the theory, that is to say, the Feynman integrals

$$\langle \mathcal{O}_1(p_1) \cdots \mathcal{O}_r(p_r) \rangle = \int_{\Sigma, \phi, \psi} \prod_i \mathcal{O}_i((\phi, \psi)(p_i)) e^{-S(\phi, \psi)} \, d\Sigma \, d\phi \, d\psi,$$

where p_i are fixed points of Σ and the $\mathcal{O}_i$ are differential forms on X, the integral over Σ denoting the integral over the complex structures of Σ up to isomorphism.

Physicists have arguments suggesting that the invariance of the action $S(\phi, \psi)$ with respect to $N = 2$-supersymmetry may be preserved at the quantum stage (that is, a central extension of the superalgebra is represented on the Hilbert space $\mathcal{H}$ in such a way that its action on observables coincides with the operator bracket on $\mathcal{H}$) precisely when the metric on X is a Kähler–Einstein metric.

This suggests associating with (X,ω) a representation of the Virasoro $N=2$-superalgebra having "central charge" $c=3n$. According to Segal [36], one may regard this representation as the infinitesimal version of a "theory of $N=2$-superconformal fields" associated with (X,ω). According to Gepner, this correspondence should be bijective (provided one considers only integer "$U(1)$ charge" representations). Mirror symmetry would be the correction that needs to be added to that statement and would correspond to the following phenomenon: the $N=2$-superalgebra admits four series of generators

$$G_r^+, \quad G_r^-, \quad \overline{G}_r^+, \quad \overline{G}_r^-, \quad r \in \mathbb{Z} + \frac{1}{2},$$

whose geometric significance depends on the interpretation of the superalgebra in terms of supersymmetric transformations: it turns out that one can construct an involution on the superalgebra acting on the odd generators as follows:

$$G_r^+ \mapsto G_r^-, \quad G_r^- \mapsto G_r^+,$$

$$\overline{G}_r^+ \mapsto \overline{G}_r^+, \quad \overline{G}_r^- \mapsto \overline{G}_r^-,$$

compatible with the Lie superbracket.

This involution has no geometric interpretation: the idea is that the representation obtained by composing the initial representation with this involution is the representation associated with the mirror (X',ω'), or again that one has the same representation with a different labeling of the generators, which must be interpreted geometrically by passing to the mirror. This yields formally the comparison of the Dolbeault cohomologies of X and X'. Indeed, following Witten [38], one can identify

$$\bigoplus_{p,q\geq 0} H^q(X, \overset{p}{\wedge} T_X)$$

with the subspace of the Hilbert space $\mathcal{H}$ formed of "chiral-chiral" states, that is to say, those annihilated by the generators

$$G_r^+, \quad \overline{G}_r^+, \quad \left(r \geq -\frac{1}{2}\right) \quad \text{and} \quad G_r^-, \quad \overline{G}_r^- \quad \left(r \geq \frac{1}{2}\right),$$

the bigrading (p,q) on the second space being furnished by the coupling of the eigenvalues of the operators J_0 and $\overline{J}_0$, where for $m \in \mathbb{Z}$ the operators J_m and $\overline{J}_m$ form a series of even generators of the superalgebra, called the *current* $U(1)$, on which the involution acts by

$$J_m \mapsto -J_m, \quad \overline{J}_m \mapsto \overline{J}_m.$$

Similarly,

$$\bigoplus_{p,q\geq 0} H^q(X, \overset{p}{\wedge} \Omega_X)$$

can be identified with the "antichiral-chiral" states, that is, annihilated by

$$G_r^-, \quad \overline{G}_r^+, \quad \left(r \geq -\frac{1}{2}\right) \quad \text{and} \quad G_r^+, \quad \overline{G}_r^- \quad \left(r \geq \frac{1}{2}\right),$$

the bigrading $(-p,q)$ on the second space being furnished by the coupling of the eigenvalues of the operators J_0, $\overline{J}_0$. By definition the chiral-chiral states of a theory are the antichiral-chiral states of the theory obtained by composition with the involution that interchanges G^+ and G^-, and the bigrading undergoes simply

the interchange $(p,q) \mapsto (-p,q)$, corresponding to the change of sign of J_0. The preceding thus provides a series of isomorphisms
$$H^q(\Omega_X^p) \cong H^q(\overset{p}{\wedge} T_{X'})$$
for the mirror (X', ω') of (X, ω).

Finally the assumptions of the theory of conformal fields (and more particularly the state/operator field correspondence) make it possible to construct a graded product on the space of chiral-chiral (resp. antichiral-chiral) states and in particular since this space is of rank 1 in bidegree (n, n) (resp. $(-n, n)$), a homogeneous form of degree n on its component of bidegree $(1, 1)$ (resp. $(-1, 1)$), which is isomorphic to $H^1(T_X)$ (resp. $H^1(\Omega_X)$).

The interpretation of these forms (physical Yukawa couplings) in terms of correlation functions of the σ-model determined by (X, ω), and the asymptotic expansion of Feynman integrals then made it possible for Witten [38] to describe the couplings obtained over $H^1(\Omega_X)$ and $H^1(T_X)$ respectively, at least for the case $n = 3$.

The former, denoted Y^ω is independent of the complex structure of X and depends instead on the parameter ω; in contrast, the latter, denoted Y^η, depends only on the complex structure of X and the choice of the holomorphic form $\eta \in H^{3,0}(X)$, whose square reflects the choice of the isomorphism $H^3(\overset{3}{\wedge} T_X) \cong \mathbb{C}$. Witten gives the following descriptions:

- the form Y^η can be identified with the composition
$$S^3 H^1(T_X) \longrightarrow H^3(\overset{3}{\wedge} T_X) \overset{\eta^2}{\cong} \mathbb{C};$$

- the form Y^ω is given by the formula
$$Y^\omega(\gamma) = \int_X \gamma^3 + \sum_{0 \neq A \in H_2(X, \mathbb{Z})} N(A) \exp\left(-\int_A \omega\right)\left(\int_A \gamma\right)^3$$

where $N(A)$ is a rational number that is an adequate substitute for the number of rational curves of class A in X, and is obtained by an integration over the set of rational curves of class A when the latter is not discrete.

Just as the mirror couples (X, ω) and (X', ω') have by definition the same associated conformal theories, their Yukawa couplings will be identified via the isomorphisms
$$H^1(T_X) \cong H^1(\Omega_{X'}), \quad H^1(T_{X'}) \cong H^1(\Omega_X)$$
by an adequate choice of η and η'.

It was in this way—and assuming that the natural choice of η' mentioned above is the correct one—that Candelas, de la Ossa, Green, and Parkes derived the identity of the two series and thus the value of the $N(d)$. Of course, the procedure assumes that the form of the mirror map has been determined a priori. That is done by the canonical coordinates and is the subtlest point of [43].

This theory gives a natural appearance to mirror symmetry that no mathematical approach has yet been able to equal.

Nevertheless, one can hardly consider it satisfactory since it rests on the hypothetical construction of the correspondence between Calabi–Yau manifolds endowed with one complexified Kähler parameter and $N = 2$-superconformal quantum field theories. It seems impossible, however, given the correctness of the predictions resulting from this approach, not to admit the existence of such a correspondence.

The problem that naturally arises mathematically, and which seems more important theoretically than mirror symmetry itself, is to give a mathematical realization of this correspondence.

Mathematical progress toward an understanding of the principle of mirror symmetry that goes beyond the construction and study of examples, resides essentially in the construction of analogous structures on the two moduli spaces $K(X)$ and $\operatorname{Def} X$ and in the formulation of mirror symmetry in terms of identification of these structures; the structures are stated and described in terms of variations of Hodge structure, which are the most subtle mathematical object associated with a deformation of complex structure.

The first analogy between these two moduli spaces is the following. We recall that $K(X)$ is open in $H^2(X, \mathbb{C})/2i\pi H^2(X, \mathbb{Z})$, and thus admits a natural flat structure, that is, the assignment of a local coordinate system, defined up to affine transformations. A first easy result is the existence of such a structure on $\operatorname{Def} X$ depending on a partial marking of the cohomology $H^n(X, \mathbb{Z})$, and given essentially by the periods of a generator of $H^{n,0}(X)$ on certain homology classes of X, which give coordinates on $\operatorname{Def} X$. The mirror map between $K(X)$ and $\operatorname{Def} X$ is thus practically determined by the condition of compatibility with these flat structures.

As mentioned above, for a Kähler manifold X there exists a direct-sum decomposition of each of the cohomology groups $H^k(X, \mathbb{C}) \cong H^k(X, \mathbb{C}) \otimes \mathbb{C}$:

$$H^k(X, \mathbb{C}) = \bigoplus_{p+q=k} H^{p,q}(X)$$

provided by the Hodge theory and satisfying a certain number of conditions. For example, $H^{p,q}(X)$ is the complex conjugate of $H^{q,p}(X)$ and, for $k = n = \dim X$, $H^{p,q}(X)$ is orthogonal to $H^{p',q'}(X)$ relative to the intersection form $\langle \, , \, \rangle$ of $H^n(X)$ for $(p', q') \neq (n - p, n - q)$. If the integer k is fixed, the local (or marked) period map of X associates with the Hodge decomposition of X_t on the fixed vector space $H^k(X, \mathbb{C})$ a deformation X_t of X accompanied by a $\mathcal{C}^\infty$ diffeomorphism of X_t with X inducing an isomorphism $H^k(X_t) \cong H^k(X)$.

It is pleasanter to consider the variation corresponding to the Hodge filtration

$$F^i H^k(X_t) = \bigoplus_{p \geq i} H^{p,k-p}(X_t),$$

which enjoys the following two remarkable properties:
- $F^i H^k(X_t)$ varies holomorphically as a function of t;
- the differential of the period map satisfies the Griffiths "transversality" condition

$$\frac{d}{dt}\left(F^i H^k(X_t)\right) \subset F^{i-1} H^k(X_t).$$

The period map is often one-to-one even for Calabi–Yau manifolds; it can be shown that it is immersive for $k = n$.

The problem is that, precisely because of the transversality condition, which is generally a non-trivial differential equation, the period map is almost never surjective, so that it can only rarely be used as it is to describe the moduli space of deformations of the complex structure of a manifold X.

In the case of Calabi–Yau manifolds, it would be very interesting to understand the connection between the period map and the construction of the conformal field theory proposed by the physicists. It is not clear that the latter must determine the former, but there is no doubt that a relation exists, since as was noted above, the spaces $H^{p,q}(X)$ are calculable from the point of view of the superconformal theory associated with X.

Another, more precise, connection is the fact that the Yukawa couplings on the tangent space $H^1(T_X)$ to $\operatorname{Def} X$ calculated by the physicists as correlation functions have a very simple interpretation in terms of the variation of Hodge structure: the form η that normalizes them can be continued as a section of the bundle with fiber $H^{n,0}(X_t)$ at the point $t \in \operatorname{Def} X$; for n vector fields $u_1, \ldots, u_n$ on $\operatorname{Def} X$ we therefore set

$$Y^\eta(u_1, \ldots, u_n) = \langle \eta_{,u_1}(\cdots(_{u_n}(\eta))\cdots)\rangle$$

where the derivatives $_{u_i}(\eta)$ are taken by regarding η as a function with values in the constant space $H^n(X, \mathbb{C})$. The polarization and transversality conditions imply easily that this does indeed define a symmetric homogeneous form of degree n on the tangent space to $\operatorname{Def} X$ at the point X. Griffiths has in addition shown that this form can be identified with the product

$$S^n H^1(T_X) \longrightarrow H^n(\overset{n}{\wedge} T_X) \overset{\eta^2}{\cong} \mathbb{C}$$

as the Yukawa coupling of the physicists. Thus we now have the following simple result, which will be stated only for the case $n = 3$, which is the one most often considered in the present book.

In the natural coordinates z_i on $\operatorname{Def} X$ derived from a partial marking of $H^3(X, \mathbb{Z})$ and for a natural choice of $\eta(z)$, a section of the bundle ($\mathcal{H}^{3,0}$ with fiber $H^{3,0}(X_z)$ at the point z) derived from the same marking, the form Y^η depends on a potential, that is, there exists a function $F(z)$ such that

$$\frac{\partial^3 F}{\partial z_i \partial z_j \partial z_k} = Y^\eta\left(\frac{\partial}{\partial z_i}, \frac{\partial}{\partial z_j}, \frac{\partial}{\partial z_k}\right).$$

The most remarkable fact in favor of mirror symmetry is the existence of analogous entities on the complexified Kähler cone $K(X)$. The point is the use of Gromov–Witten invariants of the symplectic manifold underlying X to construct a potential whose derivatives will make it possible to construct a variation of complex Hodge structure parameterized by $K(X)$. The Gromov–Witten invariants [**76**], [**78**], [**83**] are multilinear forms on the cohomology $H^*(X)$ of a symplectic manifold (X, ω) depending on the choice of a class $A \in H_2(X, \mathbb{Z})$

$$\Phi_{A,m} : H^*(X, \mathbb{Q})^{\otimes m} \longrightarrow \mathbb{Q}.$$

In short, $\Phi_{A,m}(\beta_1 \otimes \cdots \otimes \beta_m)$ is obtained by integrating the class

$$\operatorname{pr}_1^* \beta_\wedge \cdots \wedge \operatorname{pr}_m^* \beta_m \in H^*(X^m)$$

over the image of the evaluation map

$$\operatorname{ev}_m : W_{A,J,\nu} \times (\mathbb{P}^1)^{(m-3)} \to X^m$$
$$(\phi, x_1, \ldots, x_{m-3}) \mapsto (\phi(0), \phi(1), \phi(\infty), \phi(x_1), \ldots, \phi(x_{m-3})).$$

Here $W_{A,J,\nu}$ is the set of maps $\phi : \mathbb{P}^1 \to X$ such that $\phi_*([\mathbb{P}^1]) = A$ satisfying the inhomogeneous Cauchy–Riemann equation

$$\bar{\partial}_J \phi = (\operatorname{Id}, \phi)^* \nu$$

for a generic choice of pseudocomplex structure J on X compatible with the symplectic form ω and having parameter on $\mathbb{P}^1 \times X$ equal to
$$\nu \in \mathcal{C}^\infty\left(\mathrm{pr}_1^*(\Omega_{\mathbb{P}^1}^{0,1}) \otimes \mathrm{pr}_2^*(T_{X,J}^{1,0})\right).$$

The Gromov–Witten potential is then the function defined on (a conjecturally non-empty open set) $H^{\mathrm{even}}(X) = \bigoplus_i H^{2i}(X, \mathbb{C})$ by the series
$$\Psi_\omega(\gamma) = \sum_{\substack{A \in H_2(X,\mathbb{Z}) \\ m \geq 3}} \frac{1}{m!} \Phi_{A,m}(\gamma^{\otimes m}) \exp\left(-\int_A \omega\right).$$

Since X is a Calabi–Yau threefold, the potential Ψ_ω changes with ω only by translation, modulo a quadratic term in γ:
$$\Psi_{\omega'}(\gamma) = \Psi_\omega(\gamma - \omega' + \omega) + Q(\gamma), \quad \deg Q = 2.$$

It is this potential restricted to $H^2(X, \mathbb{C})$ that mirror symmetry identifies, in canonical coordinates and modulo a quadratic function, with the potential described above, from which the Yukawa couplings of the variation of Hodge structure of the mirror X' derive when X is a Calabi–Yau variety. Indeed, if we assume $n = 3$ for simplicity, this potential makes it possible to construct a variation of complex Hodge structure parameterized by $K(X)$ (assuming of convergence) as follows. The cubic derivatives of Ψ_ω make it possible to construct at each point $\gamma \in H^{\mathrm{even}}(X)$ where the series converges a product "$\bullet_\gamma$" on $H^{\mathrm{even}}(X)$ by the formula
$$\langle e_i \bullet_\gamma e_j, e_k \rangle = \frac{\partial^3 \Psi_\omega}{\partial t_i \partial t_j \partial t_k}$$
where e_i is a basis of $H^{\mathrm{even}}(X)$ and t_i are the linear coordinates corresponding to $H^{\mathrm{even}}(X)$.

An essential property of the Gromov–Witten invariants, which holds for all symplectic manifolds and is trivially satisfied in the case of Calabi–Yau manifolds threefolds, is the associativity of the product "$\bullet_\gamma$". From this one deduces, following Dubrovin [75], that the connection ∇ defined on the (trivial) tangent bundle of the manifold $H^{\mathrm{even}}(X)$ by
$$\nabla_u(\sigma)(\gamma) = d_u(\sigma)(\gamma) + u \bullet_\gamma \sigma$$
(d being the trivial connection and u, σ tangent fields) is integrable.

Finally, in the case of Calabi–Yau threefolds, the product "$\bullet_\gamma$" is compatible with the grading $H^{\mathrm{even}}(X) = \bigoplus_i H^{2i}(X)$ for all γ. From this it can be deduced that the restriction of ∇ to $H^2(X, \mathbb{C})$ is an integrable connection on the trivial bundle with fiber $H^{\mathrm{even}}(X)$ satisfying the transversality condition
$$\nabla F^i H^{\mathrm{even}}(X) \subset F^{i-1} H^{\mathrm{even}}(X) \otimes \Omega_{H^2(X)}$$
for the filtration
$$F^3 H^{\mathrm{even}} = H^0(X), \qquad F^2 H^{\mathrm{even}} = H^0 \oplus H^2,$$
$$F^1 H^{\mathrm{even}} = H^0(X) \oplus H^2 \oplus H^4, \quad F^0 H^{\mathrm{even}} = H^{\mathrm{even}}(X),$$
which provides the stated complex variation of the Hodge structure.

At this stage one can formulate mirror symmetry by conjecturing that this variation of Hodge structure is that of the mirror family $\{X'\}$ via the mirror map
$$K(X) \longrightarrow \mathrm{Def}\, X'.$$

The problem is that there is no reason to believe that it has a geometric origin, that is, that it corresponds to the variation of Hodge structure of a family of manifolds parameterized by $K(X)$.

We note also that, even in the case described above of quintics of $\mathbb{P}^4$, the equality of the two potentials described above or, what amounts to the same thing, the two power series considered in [**43**], remains a conjecture.[1]

Nevertheless, Givental has given a justification of it (unfortunately not sufficient from the point of view of rigor, due to the use of infinite products, which are assumed to have a sense in the "equivariant Floer cohomology" but whose status is uncertain) very attractive theoretically and completely independent of that of the physicists. Givental notes first of all the existence of a $\mathcal{D}$-module structure on the S^1-equivariant cohomology of a symplectic manifold $\widetilde{M}$ under the following conditions: $\widetilde{M}$ is a covering with group $\mathbb{Z}$ of a symplectic manifold M endowed with a locally Hamiltonian action of S^1, on which this action becomes globally Hamiltonian, the Hamiltonian function H satisfying

$$q^*H = H + 1$$

for $q : \widetilde{M} \to \widetilde{M}$ generating the action of $\mathbb{Z}$. On $\widetilde{M}$ the symplectic form ω then extends to an equivariant form $\tilde{\omega}$ whose class satisfies

$$q^*\tilde{\omega} = \tilde{\omega} - h$$

where h is the standard (point) generator of $H^*_{S^1}$; if p is the operator of multiplication by the class of $\tilde{\omega}$ in $H^*_{S^1}(\widetilde{M})$, we thus have the relation

$$p \circ q^* - q^* \circ p = hq^*,$$

which, if we agree to treat h as a scalar, provides the required $\mathcal{D}$-module structure if we place p in correspondence with $h\partial/\partial t$ and q^* with multiplication by e^t.

Givental applies this construction to the case when M is the loop space of $\mathbb{P}^4$ and $\widetilde{M}$ its universal covering, the action of S^1 being given by the rotation of loops. In fact, he approximates this covering by the projective variety M_∞, the direct limit of the space of Laurent polynomials

$$M_k = \mathbb{P}\Big(\Big\{\sum_{i=-k}^{k} \phi_i z^i, \phi_i \in \mathbb{C}^5\Big\}\Big),$$

the action of $\mathbb{Z}$ on M_∞ being generated by multiplication by z. The S^1-equivariant cohomology of M_∞ has thus, as above, a $\mathcal{D}$-module structure.

Givental then constructs a class that is presented as an infinite product in $H^*_{S^1}(M_\infty)$, but whose geometric interpretation in terms of counting rational curves on a general quintic of $\mathbb{P}^4$ is clear. It satisfies formally, for this $\mathcal{D}$-module structure, a differential equation that is none other than the Picard–Fuchs equation of the mirror family $\{\overline{X_\lambda/H}\}$ described above. (This differential equation has meromorphic functions of t as coefficients, and its solutions are precisely the periods of the type (3,0) form on which it depends, on locally constant homology cycles. As explained above, these periods are essential in the calculation of canonical coordinates and the normalization of Yukawa couplings on which the construction of the mirror map is based.)

---

[1]It is now proved in [**G**].

The text ends with a description of this construction, which is rather miraculous, but undoubtedly involves very intimately the phenomenon of mirroring with its central object, the symplectic geometry of the loop space, which is also the object studied by the physicists through string theory and the σ-model.

Organization of the text

The first chapter contains some results on algebraic or Kähler geometry concerning Calabi–Yau manifolds: the existence of Kähler–Einstein metrics, the smoothness of the Kuranishi family, and some specific properties of these manifolds from the point of view of Hodge theory.

Chapter 2 is devoted to a description of mirror symmetry as it emerges from "physics." On the one hand the ideas involved in conformal field theory appear to be of great scientific import; on the other hand, as explained in this introduction, there is not at present enough mathematical intuition on the mirror phenomenon to replace the theory of the physicists.

That is the motive for this chapter, whose inspiration is somewhat special, and which is not necessary for understanding the other chapters.

The other four chapters are devoted to the mathematical aspects of mirror symmetry mentioned in the introduction.

Chapter 3 explains, following Morrison in part, the outline of the calculation performed in [43], introducing the necessary notions of Hodge theory: Yukawa coupling, Picard–Fuchs equations, the monodromy theorem, Griffith-style description of cohomology and Hodge filtration of a hypersurface by residues.

Chapter 4 introduces toric geometry and is devoted to the Batyrev construction. Without entering into the most recent developments, we also explain, following Danilov and Khovanskii, the calculation of Hodge numbers of hypersurfaces of toric varieties, used by Batyrev to verify in part by means of these examples the predictions concerning the comparison of the Hodge numbers of a manifold and its mirror.

The last two chapters are connected by the introduction of Floer cohomology, which, though not necessary here, seems important given that it treats quantum cohomology from the point of view of symplectic geometry of the loop spaces.

Chapter 5 is devoted to quantum cohomology: here we describe Gromov–Witten invariants and their crucial property of "splitting," and we explain the formal consequences of this property, such as associativity of the quantic product and the partial differential equations known as the WDVV equation (Witten, Dijgraaf, Verlinde, Verlinde) satisfied by the Gromov–Witten potential.

Chapter 6 begins with an introduction to Floer cohomology, and a schematic description of the proof of the comparison theorem between Floer cohomology and the usual cohomology tensorized by the Novikov ring. This section has as its main purpose to justify the infinite products of Givental by explaining their natural interpretation (in homology) as "fundamental S^1-equivariant Floer homology classes." This chapter is also an introduction to S^1-equivariant cohomology, which is the only mathematical tool used in the Givental construction.

Acknowledgment

I wish to thank the organizers of the semester, A. Beauville, D. Eisenbud, J. Le Potier, and C. Peskine, for offering me the opportunity to teach under such ideal conditions. I am also grateful to those who took the course for the help that their questions and remarks gave me in the publication of this text. Finally, I thank Krzysztof Gawedzki, who gave a very constructive criticism of a preliminary version of the second chapter, the referees for the corrections and improvements they suggested, and Michèle Audin for close reading and for criticism and encouragement.

Note added in Translation

Since the appearance of these notes the subject has developed in several directions.

– The subject of Gromov–Witten invariants has been thoroughly investigated: they have been constructed in a purely algebraic way ([**BM**], [**BF**]); formulas for their transformations under blowing up have been found [**LR**].

– Concerning the general understanding of mirror symmetry, a new and very promising avenue has been opened in [**SYZ**]. The authors propose there to construct the mirror by dualizing special Lagrangian torus fibrations on Calabi–Yau manifolds. Some related results have been obtained in [**H**] and [**G**]. Also, Barannikov and Kontsevich [**BK**] have found a beautiful way of fomulating mirror symmetry in dimensions larger than 3, consistent with the ideas of [**12**]; they construct the "thickened" moduli space for a Calabi–Yau manifold X (which had also been discovered by Ran) and show that it has a natural Frobenius structure (see Chapter 5), namely a flat structure, a flat (complex) metric, and a cubic form on the tangent space that depends on a potential and together with the metric defines an associative algebra structure on the tangent space. Mirror symmetry can then be formulated as the identification of this moduli space with $M^*(\widehat{X}, \mathbb{C})$, where $\widehat{X}$ is the mirror, together with its Frobenius structure given by the Gromov–Witten invariants.

– Finally, the most important progress in mirror symmetry, which is not covered in these notes, is the splendid proof by Givental of the mirror symmetry conjecture for Calabi–Yau complete intersection in toric varieties. While the heuristic and mysterious computation described here (Chapter 6) remains perhaps the most striking point, Givental succeeded [**G**] in making it into a proof, the main conceptual ingredient being the introduction of the equivariant Gromov–Witten invariants, a more realistic substitute for the "equivariant Floer cohomology of the loop space."

CHAPTER 1

Calabi–Yau Manifolds

This chapter is devoted to the description of known results on the algebraic or Kähler geometry of Calabi–Yau manifolds. The two most general theorems involve precisely the two aspects of the moduli space studied by physicists: the Kähler cone, which Yau's theorem places in one-to-one correspondence with the Kähler–Einstein metrics, and the moduli space of deformations of the complex structure whose local version is the Kuranishi family, which by the Bogomolov–Tian–Todorov theorem is smooth. The proof of this last theorem is given in some detail, along with the theorems of Friedman and Kawamata–Namikawa on "smoothability," which offer the hope that new families can be constructed by smoothing singular varieties (actually those having normal crossings) with trivial canonical bundle. The chapter concludes with an introduction to the period map and the description, due to Bryant and Griffiths, of the maximal-dimensional solutions of the differential equation of transversality for Hodge structures of level 3 polarized with $h^{3,0} = 1$.

1. Yau's Theorem

1.1. Hermitian metrics. Let X be a complex manifold. The real tangent bundle $T_X^{\mathbb{R}}$ is endowed with an operator J of pseudo-complex structure satisfying $J^2 = -1$. Let h be a Hermitian metric on X, that is, the assignment of a Hermitian metric h_x on $T_X^{\mathbb{R}}$ regarded as a $\mathbb{C}$-vector space via J_x and varying differentiably with $x \in X$. The Hermitian form h can be decomposed as $h = g - i\omega$, with g and ω real and J-invariant. The conditions

(1.1) $$h(u, Jv) = -ih(u,v), \quad h(u,v) = \overline{h(v,u)}$$

can be translated as

(1.2) $$\omega(u, Jv) = g(u,v), \quad \omega(u,v) = -\omega(v,u)$$

Thus ω is a real J-invariant 2-form. This last condition is equivalent to the following: J induces a decomposition into types

(1.3) $$\overset{2}{\wedge} \Omega_X^{\mathbb{C}} = \Omega^{2,0} \oplus \Omega^{1,1} \oplus \Omega^{0,2},$$

with

$$\Omega^{2,0} = \overset{2}{\wedge} \Omega^{1,0}, \quad \Omega^{1,1} = \Omega^{1,0} \otimes \Omega^{0,1},$$

where $\Omega^{1,0} \subset \Omega_X^{\mathbb{C}}$ is the eigenspace associated with the eigenvalue i of J. We then have:

ω is J-invariant if and only if ω is a section of the bundle $\Omega^{1,1}$.

1.2. Kähler Metrics. The metric h is a *Kähler* metric if the associated form ω is closed. Its class in $H^2(X,\mathbb{R})$ is called the *Kähler class* of the metric h. When X admits a Kähler metric, Hodge theory provides a decomposition

$$H^2(X,\mathbb{C}) = H^{2,0} \oplus H^{1,1} \oplus H^{0,2}$$

given by the decomposition (1.3) for harmonic forms. The subspace $H^{1,1}$ is stable under conjugation, so that it has a real structure, and we have the following proposition:

LEMMA 1.1. *Let X be a Kähler manifold. Then the Kähler classes form an open cone in $H^{1,1}_{\mathbb{R}}(X)$.*

We have the following characterization of Kähler metrics:

LEMMA 1.2. *A Hermitian metric h is a Kähler metric if and only if the operator J is parallel under the Levi–Cività connection of $g = \operatorname{Re} h$.*

1.3. The First Chern Class. The canonical bundle of a complex manifold X is the invertible bundle of rank 1 generated by $dz_1 \wedge \cdots \wedge dz_n$ for local holomorphic coordinates z_i. Its transition functions are given by the determinants of the Jacobian matrices of the holomorphic changes of coordinates $(\partial z'_i/\partial z_j)_{(i,j)}$. If h is a Hermitian metric on (T_X, J) with matrix (h_{ij}) in the basis $\partial/\partial z_i$ for $i = 1, \ldots, n$, one can derive from it a metric h_{-K} on the anticanonical bundle $-K = \wedge^n_{\mathbb{C}} T_X$ with matrix $\det(h_{ij})$ in the base $\partial/\partial z_1 \wedge \cdots \wedge \partial/\partial z_n$. The *curvature* of h_{-K} is defined as the 2-form

$$(1.4) \qquad \omega_{-K} = \frac{1}{2i\pi} \partial\bar{\partial} \log h_{-K},$$

which is independent of the choice of the holomorphic trivialization of $-K$.

The connection with the Ricci curvature of (X,g) is the following. If (X,h) is a Kähler manifold, the Ricci curvature of (X,g) is a symmetric 2-form that is J-invariant, since J is parallel; it provides an alternating 2-form of type $(1,1)$ defined by

$$(1.5) \qquad \rho(u,v) = \operatorname{Ric}(Ju,v).$$

We then have

$$(1.6) \qquad \rho = 2\pi \omega_{-K}$$

because of the following lemma.

LEMMA 1.3. *If h is a Kähler metric, the Levi–Cività connection ∇ coincides with the Hermitian connection D on T_X.*

(This makes sense because $T_X^{1,0}$ and $T_X^{\mathbb{R}}$ are naturally identified.)
We then have

$$(1.7) \qquad \omega_{-K} = \frac{1}{2i\pi} \operatorname{Trace}(R^D)$$

where

$$R^D = D^2 \in \overset{2}{\wedge} \Omega_X \otimes \operatorname{Hom}_{\mathbb{C}}(T_X)$$

is the curvature operator of D. On the other hand, $\operatorname{Ric}(U,V)$ is the trace of the linear map $W \mapsto R^{\nabla}(U,W)V$ and the equality (1.6) then follows from Lemma 1.3, the first Bianchi identity, and the fact that $R(\cdot,\cdot)W$ is of type $(1,1)$.

DEFINITION 1.4. The *first Chern class* $c_1^{\mathbb{R}}(X) \in H^2(X, \mathbb{R})$ is the class of the 2-form ω_{-K}.

This class is in fact integral and independent of the choice of h. It can be defined more generally for any pseudocomplex structure J on X as the image in $H^2(X, \mathbb{R})$ of the Euler class of the complex bundle $\wedge_{\mathbb{C}}^n T_X$ of rank 1. It thus depends only on the deformation class of the pseudocomplex structure J.

1.4. Yau's Theorem. This theorem is the solution of a conjecture of Calabi. The most general version of it establishes the existence of Kähler metrics whose curvature is proportional to the Kähler form for manifolds whose canonical bundle is positive or zero, in the sense in which the curvature form ω_K for an adequate Hermitian metric on K is a positive or zero multiple of a Kähler form.

Such metrics are called *Kähler–Einstein* metrics.

We shall confine ourselves to the statement for the case $c_1^{\mathbb{R}}(X) = 0$:

THEOREM 1.5 (see [**23**]). *Let X be a compact Kähler manifold having the property $c_1^{\mathbb{R}}(X) = 0$, and let $\alpha \in H^2(X, \mathbb{R})$ be a Kähler class on X. Then there exists a unique Kähler metric $h = g - i\omega$ with Kähler form ω cohomologous to α and satisfying the condition $\omega_{-K} = 0$.*

2. The decomposition theorem

It will be recalled that the holonomy of a Riemannian manifold (M, g) (computed at a point m in M) is the group of linear transformations of $T_{M,m}$ obtained via parallel transport relative to the Levi–Cività connection along loops based at m. The restricted holonomy is the subgroup obtained by restricting to loops homotopic to the constant loop.

The following theorem was proved by Bogomolov [**3**] and independently by Beauville [**2**].

THEOREM 1.6. *Let X be compact Kähler manifold such that $c_1^{\mathbb{R}}(X) = 0$. Then a finite covering of X is isomorphic to a product of complex tori, compact simply connected Kähler manifolds of dimension $2m_i$ with holonomy equal to $\mathrm{Sp}(m_i)$, and compact simply connected Kähler manifolds of dimension n_j and holonomy $\mathrm{SU}(n_j)$.*

In addition to Yau's theorem, the proof uses various profound results from differential geometry:

(*i*): de Rham's theorem: *Let M be a complete simply connected Riemannian manifold that admits an orthogonal parallel decomposition*
$$T_M = \bigoplus T_M^i.$$
Then M is isomorphic, as a Riemannian manifold, to a product
$$M \cong \prod M_i \quad \text{with} \quad T_M^i \cong \mathrm{pr}_{M_i}^* T_{M_i} \subset T_M.$$

(*ii*): The classification of irreducible restricted holonomy groups due to Berger.

If M is a compact Kähler manifold with Ricci curvature zero, its restricted holonomy group is contained in $\mathrm{SU}(n)$ with $n = \dim_{\mathbb{C}} M$, and the same is true of the components M_i of the de Rham decomposition of M.

But (*ii*) implies that the irreducible restricted holonomy groups contained in SU (m_i) form the trivial group, which provides the tori that appear in the statement of Theorem 1.6 the group SU (m_i) itself, and for $m_i = 2m'_i$ the group SP (m'_i). This last is the intersection of the group SO $(2m_i)$ with the group of $\mathbb{K}$-linear automorphisms of T_{M_i,m_i}, where $\mathbb{K}$ is the group of quaternions, regarded as acting on T_{M_i,m_i} in a manner compatible with the metric, in the sense in which g_i is invariant under the three complex structures I, J, and K given by the action of $\mathbb{K}$. (We have $IJ = -JI = K$.)

The manifolds having holonomy SP (m'_i) are the holomorphic symplectic components of the de Rham decomposition, which means that they have a nowhere-degenerate holomorphic 2-form. Indeed, if I is the initial complex structure, the parallel pseudocomplex structure J derived from the parallel action of $\mathbb{K}$ on T_{M_i} provides a 2-form Ω of type $(2,0)$ (relative to I) on M_i by the formula

$$(1.8) \qquad \Omega(u,v) = h(u, Jv),$$

where h is the I-hermitian metric corresponding to g_i. As Ω is of type $(2,0)$ and parallel, hence closed, it is holomorphic and it can easily be seen that it is not degenerate. The following proposition shows finally that the holonomy controls the holomorphic forms on Kähler manifolds with Ricci curvature zero (which implies in particular the converse of the preceding assertion).

PROPOSITION 1.7 (Bochner). *Let M be a compact Kähler manifold with Ricci curvature zero; then a section of $\Omega_M^{p,0}$ is parallel if and only if it is holomorphic.*

This proposition implies that manifolds M with holonomy SU (m), where $m = \dim M$, and holonomy SP (m') satisfy the condition $h^0(\Omega_M^p) = 0$ for $0 < p < m$. (Indeed, the existence of a holomorphic form of intermediate degree restricts the holonomy group.)

Similarly, manifolds M of complex dimension $m = 2m'$ and holonomy Sp (m') satisfy the condition $h^0(\Omega_M^2) = 1$. These manifolds are also known as *hyperkählerian manifolds*, since the quaternion structure provides a continuous family of complex structures for which the initial metric g remains a Kähler metric.

Finally, this proposition shows that the global holonomy of a Kähler manifold M with Ricci curvature zero is contained in SU (n) if and only if M possesses a nowhere-vanishing holomorphic differential n-form, that is if the canonical bundle of M is trivial.

DEFINITION 1.8. The manifold X is a *Calabi–Yau manifold* if it is a compact Kähler manifold with trivial canonical bundle.

In the following chapters we shall consider only Calabi–Yau manifolds having the holonomy

$$\text{SU}(n), \quad n = \dim X > 2.$$

According to what has been said above, these manifolds satisfy the condition

$$h^0(\Omega_X^2) = 0$$

and Hodge theory then shows that

$$H^{1,1}(X) = H^2(X, \mathbb{C}).$$

The Kähler cone is thus an open set in $H^2(X, \mathbb{R})$, in accordance with Lemma 1.1. It therefore contains integer classes and the Kodaira embedding theorem shows that these manifolds are algebraic.

3. Smoothness of the local family of deformations

We shall sketch the proof of the following theorem, which is due to Bogomolov [4], [18], and Todorov, and has recently been given a new proof by Ran [15], using a more algebraic formalism.

THEOREM 1.9. *Let X be a compact Kähler manifold with trivial canonical bundle. Then the universal local family of deformations of X is smooth.*

Let J be the operator of complex structure on X. It determines a decomposition:

$$(1.9) \qquad T_X \otimes \mathbb{C} = T_X^{1,0} \oplus T_X^{0,1}.$$

A small deformation J_t of the pseudocomplex structure J determines a section α_t of $T_X^{1,0} \otimes \Omega_X^{0,1}$ as follows: to J_t there corresponds a decomposition

$$(1.10) \qquad T_X \otimes \mathbb{C} = T_{X_t}^{1,0} \oplus T_{X_t}^{0,1}.$$

Denoting by pr_1 and pr_2 the projections induced by (1.9), we set:

$$(1.11) \qquad \alpha_t = \mathrm{pr}_{1|T_{X_t}^{0,1}} \circ \left(\mathrm{pr}_{2|T_{X_t}^{0,1}}\right)^{-1} \in \mathrm{Hom}\left(T_X^{0,1}, T_X^{1,0}\right).$$

3.1. The integrability condition. By definition the vector fields of type $(0,1)$ for J_t are of the form

$$(1.12) \qquad \chi + \alpha_t(\chi), \quad \chi \in T_X^{0,1}.$$

The integrability condition for J_t can be written as

$$(1.13) \qquad [T_{X_t}^{0,1}, T_{X_t}^{0,1}] \subset T_{X_t}^{0,1},$$

that is,

$$(1.14) \qquad [\chi + \alpha_t(\chi), \chi' + \alpha_t(\chi')] \in T_{X_t}^{0,1}, \quad \forall \chi, \chi' \in T_X^{0,1}.$$

In order for the property (1.14) to hold for all fields of type $(0,1)$ on X, it suffices that it hold for the fields

$$\chi = \frac{\partial}{\partial \bar{z}_i}, \quad \chi' = \frac{\partial}{\partial \bar{z}_j},$$

corresponding to local holomorphic coordinates z_i for $i = 1, 2, \ldots, n$. But we then have

$$[\chi, \chi'] = 0,$$

$$[\chi, \alpha_t(\chi')] = \bar{\partial}_{z_i}\left(\alpha_t\left(\frac{\partial}{\partial \bar{z}_j}\right)\right) \in T_X^{1,0},$$

$$[\alpha_t(\chi'), \chi] = -\bar{\partial}_{z_j}\left(\alpha_t\left(\frac{\partial}{\partial \bar{z}_i}\right)\right) \in T_X^{1,0}.$$

Finally $[\alpha_t(\partial/\partial \bar{z}_i), \alpha_t(\partial/\partial \bar{z}_j)] \in T_X^{1,0}$ by the integrability of the pseudocomplex structure J. In order for (1.14) to hold, it therefore suffices that for all i and j:

$$(1.15) \qquad \bar{\partial}_{z_i}\left(\alpha_t\left(\frac{\partial}{\partial \bar{z}_j}\right)\right) - \bar{\partial}_{z_j}\left(\alpha_t\left(\frac{\partial}{\partial \bar{z}_i}\right)\right) = -\left[\alpha_t\left(\frac{\partial}{\partial \bar{z}_i}\right), \alpha_t\left(\frac{\partial}{\partial \bar{z}_j}\right)\right],$$

which can be compactly written as

(1.16) $$\bar{\partial}\alpha_t = -[\alpha_t, \alpha_t].$$

Now suppose that α_t can be expanded in a series of integer powers of t:

$$\alpha_t = \alpha_0 + t\alpha_1 + t^2\alpha_2 + \cdots.$$

If we assume $J_0 = J$, we have $\alpha_0 = 0$, and the equation (1.16), considered to first order, provides the condition

(1.17) $$\bar{\partial}\alpha_1 = 0.$$

We are interested in deformations of the complex structure modulo the action of the diffeomorphisms of X. It can be verified immediately that the infinitesimal action of these diffeomorphisms modifies α_1 by adjoining a term of the form $\bar{\partial}\chi$, where χ is an arbitrary section of $T_X^{1,0}$. We thus have the following proposition.

PROPOSITION 1.10. *The first-order deformations of the complex structure J are parameterized by the Dolbeault cohomology group:*

$$H^1(T_X) = \operatorname{Ker}\left(\bar{\partial}: T_X^{1,0} \otimes \Omega_X^{0,1} \to T_X^{1,0} \otimes \Omega_X^{0,2}\right) \bigg/ \operatorname{Im}\left(\bar{\partial}: T_X^{1,0} \to T_X^{1,0} \otimes \Omega_X^{0,1}\right).$$

To show the smoothness, it suffices, according to general principles due to Kodaira and Artin, to prove that for every class $\bar{\alpha}_1 \in H^1(T_X)$, there exists a formal series $t\alpha_1 + t^2\alpha_2 + \cdots$ that is a solution of (1.16), where $\alpha_1 \in T_X^{1,0} \otimes \Omega_X^{0,1}$ is $\bar{\partial}$-closed and represents $\bar{\alpha}_1$.

Suppose that we have found

$$\alpha_t^n = t\alpha_1 + \cdots + t^n\alpha_n$$

satisfying (1.16) to order n. We then seek α_{n+1} such that

$$\alpha_t^{n+1} = t\alpha_1 + \cdots + t^{n+1}\alpha_{n+1}$$

satisfies (1.16) to order $n+1$, which is equivalent to the condition

(1.18) $$\bar{\partial}\alpha_{n+1} = -\sum_{i \leq n}[\alpha_i, \alpha_{n+1-i}].$$

The second term is a $\bar{\partial}$-closed section of $T_X^{1,0} \otimes \Omega_X^{0,2}$, and we must prove that it is exact. The point is the following: if K_X is trivial, the choice of a nonzero holomorphic section of K_X provides via the interior product an isomorphism

$$\operatorname{int}: T_X^{1,0} \cong \Omega_X^{n-1}.$$

We shall show by recursion that there exists α_{n+1} satisfying Eq. (1.18) and the additional condition

$$\partial\bigl(\operatorname{int}(\alpha_{n+1})\bigr) = 0.$$

The fact that one can impose the condition $\partial\bigl(\operatorname{int}(\alpha_1)\bigr) = 0$ is an easy consequence of Hodge theory.

We now use the following fact.

3. SMOOTHNESS OF THE LOCAL FAMILY OF DEFORMATIONS

PROPOSITION 1.11 (Tian). *Let $\alpha_1, \alpha_2 \in T_X^{1,0} \otimes \Omega_X^{0,1}$. Then*
$$\mathrm{int}\,([\alpha_1, \alpha_2]) = \partial(\alpha_1 \bullet \mathrm{int}\,(\alpha_2)) - \eta\big(\partial\big(\mathrm{int}\,(\alpha_1)\big)\big) \wedge \mathrm{int}\,(\alpha_2)$$
$$+ \mathrm{int}\,(\alpha_1) \wedge \eta\big(\partial\big(\mathrm{int}\,(\alpha_2)\big)\big).$$

where "•" in the first term is given by the interior product on $T_X^{1,0}$ and the exterior product on $\Omega_X^{0,1}$, and where η is the isomorphism $K_X \otimes \Omega_X^{0,1} \cong \Omega_X^{0,1}$ given by the same section of K_X.

From this formula one can deduce that if $\partial\big(\mathrm{int}\,(\alpha_i)\big) = 0$ for $i \leq n$, the form
$$\mathrm{int}\left(-\sum_{i \leq n}[\alpha_i, \alpha_{n+1-i}]\right)$$
is ∂-exact. But this is a form of type $(n-1, 2)$ that is also $\bar\partial$-closed. Hodge theory (see [6], [21]) implies via the argument given below that $\mathrm{int}\left(-\sum_{i \leq n}[\alpha_i, \alpha_{n+1-i}]\right)$ is also $\bar\partial$-exact. It thereby follows that there exists a $(0,1)$-form α_{n+1} with values in $T_X^{0,1}$ satisfying Eq. (1.18) defined by $\mathrm{int}\,(\alpha_{n+1}) = \phi$ with

(1.19) $$\bar\partial \phi = \mathrm{int}\left(-\sum_{i \leq n}[\alpha_i, \alpha_{n+1-i}]\right).$$

To continue the recursion it remains simply to note that the form ϕ can be chosen ∂-closed, which can be seen as follows.

Let Δ be the Laplacian (for ∂ or $\bar\partial$) associated with a Kähler metric on X and which acts on differential forms of X preserving the type, and let H be the projection on the space $\mathcal{H}$ of Δ-harmonic forms. We have the following equalities

(1.20) $$\Delta = \partial\partial^* + \partial^*\partial = \bar\partial\bar\partial^* + \bar\partial^*\bar\partial$$

and the following decompositions (compatible with the complex bigrading) of the space $\bigoplus A^{\bullet,\bullet}(X)$ of complex differential forms of class $\mathcal{C}^\infty$ on X into the orthogonal direct sum:

(1.21) $$A^{\bullet,\bullet}(X) = \mathcal{H}^{\bullet,\bullet} \oplus \partial\big(A^{\bullet-1,\bullet}(X)\big) \oplus \partial^*\big(A^{\bullet+1,\bullet}(X)\big),$$

(1.22) $$A^{\bullet,\bullet}(X) = \mathcal{H}^{\bullet,\bullet} \oplus \bar\partial\big(A^{\bullet,\bullet-1}(X)\big) \oplus \bar\partial^*\big(A^{\bullet,\bullet+1}(X)\big).$$

According to Eq. (1.21) we have
$$H\left(\mathrm{int}\left(-\sum_{i \leq n}[\alpha_i, \alpha_{n+1-i}]\right)\right) = 0$$

since $\mathrm{int}\left(-\sum_{i \leq n}[\alpha_i, \alpha_{n+1-i}]\right)$ is ∂-exact. Hence, by Eq. (1.22) and the fact that this form, being $\bar\partial$-closed, is orthogonal to $\mathrm{Im}\,\bar\partial^*$,
$$\mathrm{int}\left(-\sum_{i \leq n}[\alpha_i, \alpha_{n+1-i}]\right) = \bar\partial \psi.$$

If we now expand ψ following Eq. (1.21), say $\psi = H\psi + \partial u + \partial^* v$, we obtain

(1.23) $$\bar\partial \psi = \bar\partial\partial u + \bar\partial\partial^* v.$$

But the term on the left-hand side is ∂-exact, as is $\bar\partial\partial u$, while $\bar\partial\partial^* v \in \mathrm{Im}\,\partial^*$, since the operators $\bar\partial$ and ∂^* commute. But then $\bar\partial\partial^* v = 0$ since $\mathrm{Im}\,\partial$ and $\mathrm{Im}\,\partial^*$ are orthogonal. We thus have $\bar{\partial}\psi = \bar{\partial}\partial u$, and we can take $\phi = \partial u$, which is indeed ∂-closed. $\square$

3.2. Ran's proof. This proof is more algebraic but it uses essentially the same ingredients: the isomorphism
$$\text{int}\, T_X^{1,0} \cong \Omega_X^{n-1}$$
and the degeneracy at E of the Hodge–de Rham (or Frölicher) spectral sequence.

The reasoning is as follows. Let $A_n = \text{Spec}\,\mathbb{C}[\epsilon]/\epsilon^{n+1}$, and let $\pi : \mathcal{X}_n \to A_n$ be the deformation of X to order n. From this we can derive an extension class $\eta \in H^1(T_{\mathcal{X}_{n-1}/A_{n-1}})$ corresponding to the exact sequence
$$(1.24) \qquad 0 \to T_{\mathcal{X}_{n-1}/A_{n-1}} \longrightarrow T_{\mathcal{X}_n}|_{\mathcal{X}_{n-1}} \longrightarrow \pi^*\left(T_{A_n}|_{A_{n-1}}\right) \to 0.$$

The obstruction (which lives in $H^2(T_X)$) arising in Eq. (1.18) to the construction of α_{n+1} and also to the extension of $\pi : \mathcal{X}_n \to A_n$ to $\pi : \mathcal{X}_{n+1} \to A_{n+1}$ can be simply expressed by considering the exact sequence
$$(1.25) \qquad 0 \to T_X \longrightarrow T_{\mathcal{X}_n/A_n} \longrightarrow T_{\mathcal{X}_{n-1}/A_{n-1}} \to 0.$$

The associated long exact sequence provides a connecting map
$$\delta : H^1(T_{\mathcal{X}_{n-1}/A_{n-1}}) \longrightarrow H^2(T_X).$$
It turns out that this obstruction is equal to $\delta(\eta)$. The fact that $\delta(\eta) = 0$ now results from the existence of the isomorphism
$$\text{int}_n : T_{\mathcal{X}_n/A_n} \cong \Omega_{\mathcal{X}_n/A_n}^{n-1}$$
(since K_X is trivial $K_{\mathcal{X}_n/A_n}$ is trivial) and from Hodge theory, which shows that the Hodge bundles are compatible with base changes.

4. Smoothability of Calabi–Yau manifolds with normal crossings

Let X be a manifold of dimension n with normal crossings, that is,
$$X = \bigcup_{i=1}^{N} X_i$$
with X_i being smooth or algebraic Kähler manifolds of dimension n such that all the intersections $X_{i_1} \cap \cdots \cap X_{i_k}$ are transverse. Locally X is isomorphic to the hypersurface with equation $z_1 \cdots z_r = 0$ in a neighborhood U of 0 in $\mathbb{C}^{n+1}$ for a certain $r \leq n+1$.

The problem under consideration involves the possibility of *smoothing* X, that is, finding a manifold $\mathcal{X}$ and a proper flat morphism $\pi : \mathcal{X} \longrightarrow \Delta$ such that
$$X \cong \pi^{-1}(0).$$
For $t \neq 0$ and t small, the fibers X_t are then smooth. There exists a fairly simple combinatorial condition for that, discovered by Friedman [9]. Let
$$D = \text{Sing}(X).$$
We construct a line bundle $\mathcal{O}_D(-X)$ on D by the formula
$$(1.26) \qquad \mathcal{O}_D(-X) = \mathcal{I}_{X_1}/\mathcal{I}_D \cdot \mathcal{I}_{X-1} \otimes_{\mathcal{O}_D} \cdots \otimes_{\mathcal{O}_D} \mathcal{I}_{X_n}/\mathcal{I}_D \cdot \mathcal{I}_{X_N},$$
where the sheaves of ideals $\mathcal{I}_{X_i}$ and $\mathcal{I}_D$ are defined as follows. Let $x \in X$ and let U be an open set, $x \in U \subset X$, isomorphic to a neighborhood of 0 in the hypersurface of

$\mathbb{C}^{n+1}$ defined by $z_1 \cdots z_{r(x)}$, the isomorphism mapping x to 0. Then $\mathcal{J}_D$ is generated in a neighborhood of x by the functions $z_1 \cdots \hat{z}_i \cdots z_{r(x)}$ for $i \leq r(x)$, and $\mathcal{J}_{X_i}$ is generated in a neighborhood of x by 1 if $x \notin X_i$, and by $z_{k(i)}$ otherwise, where $k(i) \leq r(x)$ is defined by the condition that $z_{k(i)}$ vanishes on the image of $X_i \cap U$.

If $X \subset Y$ is a hypersurface with Y smooth, we have

$$\mathcal{O}_D(-X) \cong \mathcal{O}_Y(-X)_{|D}.$$

From this one can deduce the following proposition.

PROPOSITION 1.12 (Friedman). *A necessary condition for smoothability of X is $\mathcal{O}_D(-X) \cong \mathcal{O}_D$.*

When this condition holds, X is said to be *d-semistable*. The manifold X with normal crossings has an invertible canonical bundle, since it is locally a complete intersection. We now assume that $K_X \cong \mathcal{O}_X$. We have the following theorem.

THEOREM 1.13 (Kawamata–Namikawa [11]). *Let X be a d-semistable manifold with normal crossings and trivial canonical bundle satisfying the following conditions:*
- $H^{n-1}(X, \mathcal{O}_X) = \{0\}$;
- $H^{n-2}(X^{[0]}, \mathcal{O}_{X^{[0]}}) = \{0\}$ *where* $X^{[0]} := \bigsqcup X_i$.

Then X is smoothable.

The idea is to introduce the notion of a *logarithmic structure* on X, that is, a covering U_λ of a neighborhood of $\text{Sing}(X)$ in X and for each λ, functions $z_1^\lambda, \ldots, z_{n+1}^\lambda$ on U_λ such that z_i^λ is invertible for $i > r(\lambda)$ and there exists an immersion $\phi : U_\lambda \to \mathbb{C}^{n+1}$ realizing U_λ as a neighborhood of zero in the hypersurface $z_1 \cdots z_{r(\lambda)} = 0$, with $z_i^\lambda = \phi^*(z_i)$ for $i \leq r(\lambda)$. We require in addition that for a certain permutation σ of $\{1, \ldots, n+1\}$ we have $z_i^\lambda = u_{\lambda\mu}^i z_{\sigma(i)}^\mu$ for $i \leq n+1$ on $U_\lambda \cap U_\mu$ with the condition "$u_{\lambda\mu}^i$ is invertible and $\prod_i u_{\lambda\mu}^i = 1$."

Such a logarithmic structure exists if and only if X is d-semistable (see [11]). We have the notion of a logarithmic deformation (as usual we denote a family of deformations by $\mathcal{X}$ and its central fiber by X).

Let $D_1, \ldots, D_m$ be the connected components of D. Let A be a local Artin $\mathbb{C}$-algebra and s_i, $i = 1, \ldots, m$, elements of the maximal ideal M_A of A. Assigning $s_i \in M_A$ is equivalent to assigning a $\mathbb{C}[[t_1, \ldots, t_m]]$-algebra structure on A. We say that

$$\pi : \mathcal{X} \longrightarrow \text{Spec}\, A$$

is a *logarithmic deformation of X parameterized by $s_1, \ldots, s_m$* if π is flat, $X \cong \pi^{-1}(0)$, and there exists an open covering U_λ of a neighborhood of D in $\mathcal{X}$ and functions $z_1^\lambda, \ldots, z_{n+1}^\lambda$ on U_λ satisfying the equation

$$\prod_i z_i^\lambda = s_i \quad \text{for} \quad U_\lambda \cap D_i \neq \emptyset.$$

We impose the same condition as above for the passage from U_λ to U_μ. These assignments are to induce the initial logarithmic structure on X.

Kawamata and Namikawa show that under the hypotheses of the theorem the logarithmic deformation functor defined on the category of local Artin $\mathbb{C}[[t_1 \ldots, t_m]]$-algebras is not obstructed, which means, roughly speaking, that it is "representable" by a smooth formal scheme over $\text{Spec}\,\mathbb{C}[[t_1, \ldots, t_m]]$. This implies the smoothability (assuming that one can pass from the formal category to the analytic category), for if we have a logarithmic analytic deformation $\mathcal{X} \to B \to V$ over an open neighborhood V of 0 in $\mathbb{C}^m$ for which the second map is a submersion, a point $s \in B$ over a point $t \in V$ sufficiently close to zero and satisfying $t_i \neq 0$ for all $i \leq m$ parameterizes a smooth fiber X_s.

The proof of smoothness is very similar to that of Ran. A logarithmic structure on the manifold X with normal crossings makes it possible to construct a complex of logarithmic forms $\Omega_X^\bullet(\log)$ satisfying the condition

$$\tag{1.27} \overset{n}{\wedge}\left(\Omega_X(\log)\right) \cong K_X \cong \mathcal{O}_X.$$

With the preceding notation $\Omega_X(\log)$ is the free $\mathcal{O}_X$-module generated in the open sets U_λ by Ω_{U_λ}, and by the dz_i^λ/z_i^λ for $i \leq n+1$, with the relation

$$\sum_{i=1}^{n+1} \frac{dz_i^\lambda}{z_i^\lambda} = 0.$$

(On $U = X - \text{Sing}(X)$ it is simply equal to Ω_U.) We set

$$\Omega_X^k(\log) = \overset{k}{\wedge}\Omega_X(\log).$$

These complexes have a Hodge–de Rham spectral sequence that degenerates at E_1. On the other hand the first-order logarithmic deformations are parameterized by $H^1\bigl(T_X(\log)\bigr)$, where $T_X(\log)$ is the dual bundle to $\Omega_X(\log)$ with obstruction in $H^2\bigl(T_X(\log)\bigr)$. The triviality of K_X gives an isomorphism

$$\tag{1.28} T_X(\log) \cong \Omega_X^{n-1}(\log).$$

As in Ran's proof, the essential point in obtaining the smoothness is then to prove that for a logarithmic deformation $\pi : \mathcal{X} \to B$ of X the sheaf

$$R^1\pi_* T_{\mathcal{X}/B}(\log)$$

is locally free over B, where $T_{\mathcal{X}/B}(\log) \cong \bigl(\Omega_{\mathcal{X}/B}(\log)\bigr)^*$ is the relative version of the sheaves described above. But this is a consequence of the isomorphism

$$T_{\mathcal{X}/B}(\log) \cong \Omega_{\mathcal{X}/B}^{n-1}(\log)$$

and the following proposition, due to Steenbrink, which in turn is a consequence of the mixed Hodge theory.

PROPOSITION 1.14. *If $\pi : \mathcal{X} \to B$ is a logarithmic deformation of X, the sheaves $R^l\pi_*\Omega_{\mathcal{X}/B}(\log)$ are locally free over B.*

5. The period map

5.1. The local system and Hodge filtrations. Let $\pi : X \to B$ be a proper smooth family with projective or Kähler fibers of dimension n. For each $b \in B$ we use the notation

$$X_b := \pi^{-1}(b).$$

The direct image $R^n\pi_*\mathbb{Z}$ is a local system on B. Locally $R^n\pi_*\mathbb{Z} \cong H^n(X_0, \mathbb{Z})$, given by a $\mathcal{C}^\infty$ trivialization $\mathcal{X} \cong X_0 \times B$.

We denote the holomorphic fiber $R^n\pi_*\mathbb{Z} \otimes \mathcal{O}_B$ by $\mathcal{H}^n$.

It is endowed with the integrable Gauss–Manin connection ∇, for which the parallel sections are the sections of $R^n\pi_*(\mathbb{C})$.

On $H^n(X_b, \mathbb{C})$ we have the Hodge decomposition

(1.29) $$H^n(X_b, \mathbb{C}) = \bigoplus_{p+q=n} H^{p,q}(X_b)$$

with $H^{p,q}(X_b) \cong H^q(X_b, \Omega^p_{X_b})$. Let

$$F^p H^n(X) = \bigoplus_{r \geq p} H^{r,n-r}(X).$$

We then have the following theorem.

THEOREM 1.15 (Griffiths [10]). *Assigning $F^p H^n(X_b) \subset H^n(X_b, \mathbb{C}) = \mathcal{H}^n_{(b)}$ for all b in B defines a holomorphic subbundle $F^p \mathcal{H}^n \subset \mathcal{H}^n$.*

We note that the rank of $F^p H^n(X_b)$ is independent of b, the argument being that the numbers $h^{p,q}(X_b)$ are lower semicontinuous on B, while their sum for $p + q = k$ is constant and equal to the Betti number $b_k(X_B)$. This suffices to imply that the subspace $F^p H^n(X_b) \subset H^n(X_b, \mathbb{C})$ varies in a $\mathcal{C}^\infty$ manner according to Hodge theory (see [6]).

The proof of Griffiths is the following. We note first of all that one can obtain a $\mathcal{C}^\infty$ trivialization of the family $\mathcal{X}$ over B

$$\mathcal{X} \cong X_0 \times B$$

for a small neighborhood B of 0, so that the fibers of the induced map $\mathrm{pr}_1 : \mathcal{X} \to X_0$ are complex submanifolds of $\mathcal{X}$ (this statement is proved, for example, in the second part of [10]).

If $\eta = (\eta_b)_{b \in B}$ is a $\mathcal{C}^\infty$-section of the bundle with fiber $F^p H^n(X_b)$, one can, by Hodge theory, represent it by a family $\tilde{\eta}_b$ of differential forms on the fibers, varying in a $\mathcal{C}^\infty$ manner, and such that $\tilde{\eta}_b$ is of type $(n, 0) + \cdots + (p, n-p)$ on X_b.

The hypothesis made about the trivialization then implies that the form $\tilde{\eta}$ induced on $\mathcal{X}$ having restriction $\tilde{\eta}_b$ on X_b and having zero interior product with horizontal tangent vectors is of type $(n, 0) + \cdots + (p, n-p)$ on $\mathcal{X}$. The theorem then follows from the Cartan–Lie formula, which computes $\nabla_\chi \eta(0)$ when χ is a field tangent to B as

(1.30) $$\nabla_\chi \eta(0) = \text{class of } \mathrm{int}_{\tilde{\chi}}(d\tilde{\eta})_{|X_0} \text{ in } H^n(X_0, \mathbb{C}),$$

where $\tilde{\chi}$ is the horizontal lifting of χ over $\mathcal{X}$. Indeed, if χ is of type $(0, 1)$ on B, then $\tilde{\chi}$ is of type $(0, 1)$ on $\mathcal{X}$ because of the hypothesis made on the trivialization, so that $\mathrm{int}_{\tilde{\chi}}(d\tilde{\eta})_{|X_0}$ is a (closed) form of type $(n, 0) + \cdots + (p, n-p)$, and hence of class in $F^p H^n(X_0)$.

A more algebraic way to see this theorem is to introduce the relative holomorphic de Rham complex

$$\Omega^\bullet_{\mathcal{X}/B} = \bigoplus \overset{k}{\wedge} \Omega_{\mathcal{X}/B}$$

endowed with the vertical differential. We have

$$\mathbb{R}^n \pi_*(\Omega^\bullet_{\mathcal{X}/B}) = \mathcal{H}^n,$$

since this complex is a resolution of $\pi^{-1}(\mathcal{O}_B)$. We then have

(1.31) $$F^p \mathcal{H}^n = \mathbb{R}^n \pi_* \big(0 \to \overset{p}{\wedge} \Omega_{\mathcal{X}/B} \to \cdots \to \overset{n}{\wedge} \Omega_{\mathcal{X}/B} \to 0\big),$$

which makes the holomorphic character of the Hodge filtration obvious. $\square$

In addition we have the following theorem.

THEOREM 1.16 (Griffiths [10]). *The Gauss–Manin connection satisfies the following transversality condition:*
$$\nabla F^p \mathcal{H}^n \subset F^{p-1}\mathcal{H}^n \otimes \Omega_B.$$

If we retain the construction and notations of the Griffiths-style proof of Theorem 1.15, this follows again from formula (1.30). We now take χ of type $(1,0)$ on B. Since $\tilde{\eta}$ is of type $(n,0) + \cdots + (p, n-p)$ on $\mathcal{X}$, the form $\mathrm{int}_{\tilde{\chi}}(d\tilde{\eta})|_{X_0}$ is of type $(n,0) + \cdots + (p-1, n-p+1)$ on X_0, which shows that $\nabla_\chi \eta(0)$ is indeed in $F^{p-1}H^n(X_0)$.

As above, this theorem can also be viewed more algebraically by noting that ∇ can be constructed as follows. We have the exact sequence
$$(1.32) \qquad 0 \to \Omega^{\bullet-1}_{\mathcal{X}/B} \otimes \pi^*\Omega_B \longrightarrow \Omega^{\bullet}_{\mathcal{X}}/\pi^*\Omega^2_B \wedge \Omega^{\bullet-2}_{\mathcal{X}} \longrightarrow \Omega^{\bullet}_{\mathcal{X}/B} \to 0$$
and ∇ can be obtained simply as the map $\mathbb{R}^n \pi_*(\Omega^{\bullet}_{\mathcal{X}/B}) \to \mathbb{R}^n \pi_*(\Omega^{\bullet}_{\mathcal{X}/B}) \otimes \Omega_B$ provided by the long exact sequence associated with (1.32). □

5.2. The differential of the period map. The period map $\mathcal{P}$ associates with an element $b \in B$ the flag
$$F^p H^n(X_B) \subset H^n(X_b, \mathbb{C}) \cong H^n(X_0, \mathbb{C}).$$

Its differential is thus described by a series of maps
$$\phi_p : T_{B,b} \longrightarrow \mathrm{Hom}\left(F^p H^n(X_b), H^n(X_b, \mathbb{C})/F^p H^n(X_b)\right)$$
satisfying the obvious compatibility conditions
$$\phi_p|_{F^{p+1}H^n(X_b)} = \phi_{p+1} \mod F^p H^n(X_b).$$
In terms of the connection ϕ_p is obtained using the composition
$$(1.33) \qquad F^p \mathcal{H}^n \subset \mathcal{H}^n \xrightarrow{\nabla} \mathcal{H}^n \otimes \Omega_B \longrightarrow \mathcal{H}^n/F^p\mathcal{H}^n \otimes \Omega_B,$$
which is an $\mathcal{O}_B$-linear map. By transversality we find that $\mathrm{Im}\,\phi_p(T_{B(b)})$ can be factored over the quotient $(F^p\mathcal{H}^n/F^{p+1}\mathcal{H}^n)_b = H^{n-p}(\Omega^p_{X_b})$ and that it has values in $(F^{p-1}\mathcal{H}^n/F^p\mathcal{H}^n)_b = H^{n-p+1}(\Omega^{p-1}_{X_b})$. We then have,

THEOREM 1.17 (Griffiths [10]). *The arrow*
$$\phi_p : T_{B(b)} \longrightarrow \mathrm{Hom}\left(H^{n-p}(\Omega^p_{X_b}), H^{n-p+1}(\Omega^{p-1}_{X_b})\right)$$
constructed above is equal to the composition
$$(1.34) \qquad T_{B(b)} \xrightarrow{\rho} H^1(T_{X_b}) \xrightarrow{\mathrm{prod.int.}} \mathrm{Hom}\left(H^{n-p}(\Omega^p_{X_b}), H^{n-p+1}(\Omega^{p-1}_{X_b})\right)$$
where ρ is the Kodaira–Spencer map that classifies the infinitesimal deformations.

In the proof given by Griffiths this can be seen by noting that the Kodaira–Spencer map (see Proposition 1.10)
$$\rho : T_{B,b} \to H^1(T_{X_b})$$
can be realized with the notation introduced in §5.1 by associating with χ a holomorphic field on B, the form $\bar\partial \widetilde{\chi}|_{X_b}$, which is a $(0,1)$-form with values in T_{X_b}, and analyzing the component of type $(p-1, n-p+1)$ of the form $\mathrm{int}_{\tilde\chi}(d\tilde\eta)|_{X_b}$.

In the algebraic version this result can be seen by remarking that ρ is the connecting arrow associated with the exact sequence

$$(1.35) \qquad 0 \to T_{X_b} \longrightarrow T_{\mathfrak{X}|X_b} \longrightarrow \pi^*(T_{B(b)}) \to 0$$

and using the description (1.32) of ∇. $\qquad\square$

COROLLARY 1.18. *Let X be a Calabi–Yau manifold of dimension n and let $\mathcal{M}$ be the local universal family of deformations of X. Then the period map $\mathcal{P}$ defined on $\mathcal{M}$ having values in a flag manifold on $H^n(X)$ is an immersion.*

In fact the first component ϕ_n (where $n = \dim X$) of $d\mathcal{P}$ is equal, according to Theorem 1.17, to the arrow

$$(1.36) \qquad H^1(T_{X_b}) \to \mathrm{Hom}\left(H^0(K_{X_b}), H^1(\Omega_{X_b}^{n-1})\right)$$

given by the interior product. As K_{X_b} is trivial, this arrow is clearly an isomorphism.

5.3. The local period domain. The Hodge decomposition $H^n = \bigoplus H^{p,q}$ is subject to additional conditions connected with the intersection form $\langle\,,\,\rangle$ on $H^n(X)$. The latter is integral, unimodular, and symmetric if n is even and skew-symmetric otherwise. We then have

(*i*): $H^{p,q} \perp H^{p',q'}$ if $p + p' \ne n$;
(*ii*): the Hermitian form

$$(\eta, \eta') = (-1)^{n(n-1)/2} i^{p-q} \langle \eta, \bar{\eta}' \rangle$$

is positive definite on $H^{p,q}_{\mathrm{prim}}$ when $p + q = n$.

It is assumed here that X is algebraic and endowed with an integer Kähler class ω. We then have the strong Lefschetz theorem, which asserts that

$$(1.37) \qquad \omega^{n-k} : H^k(X, \mathbb{Q}) \to H^{2n-k}(X, \mathbb{Q}),$$

is an isomorphism (of Hodge structures). The Lefschetz decomposition that results is

$$(1.38) \qquad H^s(X, \mathbb{Q}) = \bigoplus_{2r \le s} \omega^r \wedge H^{s-2r}(X, \mathbb{Q})_{\mathrm{prim}}$$

where

$$H^k(X, \mathbb{Q})_{\mathrm{prim}} := \mathrm{Ker}\,(\omega^{n-k+1}) \subset H^k(X, \mathbb{Q}).$$

If X is a Calabi–Yau threefold with $H^1(X) = \{0\}$, we have

$$H^3(X) = H^3(X)_{\mathrm{prim}}.$$

After fixing a polarization, that is, a class ω such as above (which is independent of b and must therefore be a Kähler class on X_b for every b), we work with the polarized period map, that is, with the variation of Hodge structure on H^n_{prim}.

The local period domain $\mathcal{D}$ is then a set of filtrations of correct rank on $H^n(X, \mathbb{C})_{\mathrm{prim}}$ satisfying conditions (*i*) and (*ii*) above with

$$H^{p,q} = F^p H^n \cap \overline{F^q H^n}$$

and the condition
(*iii*) $F^p H^n \cap \overline{F^{q+1} H^n} = \{0\}$ if $p + q = n$.

5.4. The global period domain. If we consider families $\pi : \mathcal{X} \to B$, where B is not simply connected, we cannot trivialize $R^n \pi_* \mathbb{Z}$ globally. We have a monodromy action

$$\rho : \pi_1(B, b) \longrightarrow \operatorname{Aut}\bigl(H^n(X_b, \mathbb{Z}), \langle\,,\,\rangle\bigr) =: \Gamma \tag{1.39}$$

obtained by trivializing the family $\mathcal{X}$ in a $\mathcal{C}^\infty$ manner along paths in B. We have an obvious action $F^\bullet \mapsto \phi(F^\bullet)$ from Γ onto $\mathcal{D}$ for $\phi \in \Gamma$, and we can still define the period map

$$\mathcal{P} : B \to \mathcal{D}/\Gamma. \tag{1.40}$$

REMARK 1.19. In general $\mathcal{P}$ cannot be onto because of the transversality condition (Theorem 1.16), which implies

$$\mathcal{P}_*(T_B) \subset T_{\mathcal{D},\mathrm{hor}} := \bigoplus \operatorname{Hom}(H^{p,q}, H^{p-1,q+1}). \tag{1.41}$$

6. Calabi–Yau threefolds

Bryant and Griffiths [1] analyze the transversality condition in this case: they give the lattice $H^3(X, \mathbb{Z})$ endowed with its symplectic form $\langle\,,\,\rangle$. They consider the period map for the holomorphic forms, that is, the map,

$$b \in B \mapsto H^{3,0}(X_b) \subset H^3(X_0, \mathbb{C}).$$

Since the rank of $H^{3,0}$ is equal to 1, it is a map with values in $\mathbb{P}(H^3(X_0, \mathbb{C}))$. The Hodge filtration

$$\mathcal{H}^{3,0} \subset F^2\mathcal{H}^3 \subset F^1\mathcal{H}^3 \subset F^0\mathcal{H}^3 = H^3(X_0, \mathbb{C}) \otimes \mathcal{O}_B \tag{1.42}$$

satisfies the conditions

(a): $F^2\mathcal{H}^3_b$ is a totally isotropic maximal subspace of $H^3(X_b, \mathbb{C}) \cong H^3(X_0, \mathbb{C})$ (see 1.5.3 (i)).
(b): $d(\mathcal{H}^{3,0}) \subset F^2\mathcal{H}^3$ by transversality (Theorem 1.16), where $d(\mathcal{H}^{3,0})$ is the subbundle of $\mathcal{H}^3$ generated over $\mathcal{O}_B$ by the derivatives of holomorphic sections of $\mathcal{H}^{3,0}$ with respect to holomorphic fields on B;
(c): more precisely, we have $d(\mathcal{H}^{3,0}) = F^2\mathcal{H}^3$ from the fact that K_{X_b} is trivial (see the proof of Corollary 1.18);
(d): $F^1\mathcal{H}^3 = \mathcal{H}^{3,0\perp}$ (see 1.5.3 (i)).

6.1. The contact structure on $\mathbb{P}(H^3(X_0, \mathbb{C}))$. We now describe the construction of Bryant and Griffiths. The symplectic form $\langle\,,\,\rangle$ endows $\mathbb{P}(H^3(X_0, \mathbb{C}))$ with a contact structure, that is, a distribution of codimension 1 defined locally by a 1-form α defined up to multiplication by a nonzero function and satisfying the condition $\alpha \wedge (d\alpha)^N \neq 0$, where $2N + 1 = \dim \mathbb{P}(H^3(X_0, \mathbb{C}))$. We have

$$T_{\mathbb{P}(H^3(X-0,\mathbb{C})),\Omega} \cong H^3(X_0, \mathbb{C})/\langle\Omega\rangle$$

and α_Ω is simply the form $\langle\Omega, \cdot\rangle$ on $H^3(X_0, \mathbb{C})/\langle\Omega\rangle$ (which indeed depends on the choice of Ω in the line $\langle\Omega\rangle \in \mathbb{P}(H^3(X-0, \mathbb{C}))$). We have

LEMMA 1.20. *Let $Z \subset \mathbb{P}(H^3(X_0, \mathbb{C}))$ be an integral manifold having contact structure (that is, such that $\alpha_{|Z} = 0$); then $\dim Z \leq N$, and for all $z \in Z$, $T_{Z,z} \subset z^\perp/\langle z\rangle$ is totally isotropic for the symplectic form induced by $\langle\,,\,\rangle$.*

Indeed, $\alpha_{|Z} = 0$ implies $d\alpha_{|Z} = 0$, and one can verify immediately that at every point $\Omega \in \mathbb{P}(H^3(X_0, \mathbb{C}))$ we have the following relation: $d\alpha_{(\Omega)|\{\alpha=0\}}$ (where $\{\alpha = 0\} \subset T_{\mathbb{P}(H^3(X_0,\mathbb{C})),\Omega}$ is the hyperplane determined by α_Ω) is equal to the symplectic form induced by $\langle\,,\,\rangle$ on $\Omega^\perp/\langle\Omega\rangle$. Since $T_{Z,\Omega} \subset \Omega^\perp/\langle\Omega\rangle$ is totally isotropic for this symplectic form, we have indeed $\dim Z \leq N$. □

Assume that we have an integral manifold Z of the contact structure of dimension N. For $z \in Z$ we set

$$(1.43) \qquad H^{3,0} = F^3 H^3 = \langle z \rangle, \quad F^2 H^3 = \langle z, T_{Z,z} \rangle, \quad F^1 H^3 = z^\perp.$$

LEMMA 1.21. *The formula (1.43) defines a variation of Hodge structure of weight 3 with $h^{3,0} = 1$ on the open set on which Z is smooth and the filtration (1.42) satisfies the polarization condition 1.5.3 (ii).*

Indeed, the polarization conditions 5.3 (i) are satisfied because of Lemma 1.20. The spaces $F^2 H^3 / F^3 H^3$ and $F^1 H^3 / F^2 H^3$ have the correct dimension N, for $\dim Z = N$. On the other hand, by definition of $F^2 H^3$, the first transversality condition $d(H^{3,0}) \subset F^2 H^3$ is satisfied, and it suffices to prove that the same is true for the second transversality condition $d(F^2 H^3) \subset F^1 H^3$. But by taking coordinates x_i on Z and a lifting $z \mapsto \tilde{z}$ with $\tilde{z} \in H^3(X_0, \mathbb{C})$, we have $\langle \tilde{z}, \partial/\partial x_i(\tilde{z}) \rangle = 0$. If we derive with respect to x_j, knowing that

$$\left\langle \frac{\partial}{\partial x_i}(\tilde{z}), \frac{\partial}{\partial x_j}(\tilde{z}) \right\rangle = 0 \quad \text{(Lemma 1.20)},$$

we obtain

$$\left\langle \tilde{z}, \frac{\partial^2}{\partial x_j \partial x_i}(\tilde{z}) \right\rangle = 0,$$

or again $d(F^2 H^3) \subset F^1 H^3$. □

The variations of Hodge structure of Calabi–Yau threefolds, and more generally variations of Hodge structure of maximal dimension, of weight 3 with $h^{3,0} = 1$ provide conversely, according to conditions (a) and (c) above, integral manifolds of the contact structure on $\mathbb{P}(H^3(X_0, \mathbb{C}))$. According to (b) and (d), any such variation of Hodge structure is necessarily constructed as in (1.43), starting from the associated integral manifold.

These integral manifolds are also easy to construct: one can construct a birational transformation that preserves the contact structures between $\mathbb{P}(V)$, for a $\mathbb{C}$-vector space V of dimension $2N + 2$ endowed with a symplectic structure, and $\mathbb{P}(\Omega_M)$, where $M = \mathbb{P}(W)$, W being a $\mathbb{C}$-vector space of dimension $N + 2$, with the contact structure on $\mathbb{P}(\Omega_M)$ being given by the tautologic 1-form. The integral manifolds of maximum dimension of the contact structure over $\mathbb{P}(\Omega_M)$ that project immersively on M are simply obtained locally starting from a hypersurface $f = 0$ in M and by associating with it $Z = \{(x, df_x), f(x) = 0\}$.

7. Examples of Calabi–Yau manifolds

7.1. Complete intersections. Let $X \subset \mathbb{P}^n$ be defined by the equations $F_1 = \cdots = F_k = 0$ with $k \leq n$ and $\deg(F_i) = d_i$. For a generic choice of F_i the manifold

X is a smooth manifold of dimension $(n-k)$; the canonical bundle of X can be calculated by the adjunction formula. The exact sequence

(1.44) $$0 \to T_X \longrightarrow T_{\mathbb{P}^n}|_X \longrightarrow \bigoplus_{i \leq k} \mathcal{O}_X(d_i) \to 0$$

shows that

$$K_X = K_{\mathbb{P}^n/X}(\Sigma d_i).$$

But $K_{\mathbb{P}^n} = \mathcal{O}_{\mathbb{P}^n}(-n-1)$, by the Euler exact sequence:

(1.45) $$0 \to \mathcal{O}_{\mathbb{P}^n} \longrightarrow H^0(\mathcal{O}_{\mathbb{P}^n}(1))^* \otimes \mathcal{O}_{\mathbb{P}^n}(1) \longrightarrow T_{\mathbb{P}^n} \to 0.$$

Thus K_X is trivial when $\sum d_i = n+1$. By Lefschetz' theorem these manifolds satisfy

$$H^0(\Omega_X^i) = \{0\} \quad \text{for} \quad 0 < i < n-k.$$

When $n - k \neq 2$, it can be shown that one thereby constructs a complete family of deformations. One can also do this construction in inhomogeneous projective spaces with isolated singularities and more generally on Fano toric varieties (see Chapter 4).

7.2. Quotients and desingularizations. This construction is particularly valuable in dimension 3 (the point is to desingularize while preserving the condition $K_X \cong \mathcal{O}_X$; we shall confine ourselves to the following example. Let X be a Calabi–Yau threefold, i an involution on X acting trivially on the type (3,0) form of X. The fixed points of i then form a union of smooth disjoint curves C_k. We then have the following lemma.

LEMMA 1.22. *The blow-up $\widetilde{X/i}$ of X/i along $\sqcup C_k$ is a smooth manifold with trivial canonical fiber.*

In fact $\widetilde{X/i}$ is also the quotient of $\tilde{X}$ (blow-up of X along $\sqcup C_k$) by the lifting $\tilde{i}$ of i. The involution $\tilde{i}$ has as fixed points the disjoint union of exceptional divisors E_k over C_k. We have, on $\tilde{X}$,

(1.46) $$K_{\tilde{X}} = \tau^*(K_X) + \sum E_k = \sum E_k = r^* K_{\widetilde{X/i}} + \sum E_k$$

where τ is the extension map and r is the quotient map. Thus $r^* K_{\widetilde{X/i}}$ is trivial and $K_{\widetilde{X/i}}$ is a torsion point in $\text{Pic}(\widetilde{X/i})$. On the other hand, this bundle has a nonzero section, since the type (3,0) form of $\tilde{X}$ is invariant under $\tilde{i}$. Thus $K_{\widetilde{X/i}}$ is trivial. □

More generally, Roan [**16**] has constructed a desingularization with trivial canonical bundle for quotients of Calabi–Yau threefolds by an Abelian group preserving the type (3,0) form (this result was likewise obtained by Markushevich [**13**]).

8. Mirrors

Let X be a Calabi–Yau manifold of dimension $n \neq 2$ satisfying (see 1.2):

$$h^i(\mathcal{O}_X) = 0, \quad 0 < i < n.$$

We consider the set $\mathcal{M}_X$ of isomorphism classes of a complex structure X_t on X (obtained by deformation of the initial complex structure) and assignment of a "complexified Kähler parameter" on X_t, that is to say, a class

$$\omega = \alpha + i\beta \text{ defined modulo } 2\pi i H^2(X_t, \mathbb{Z}),$$

where $\beta \in H^2(X_t, \mathbb{R})$ is defined modulo $2\pi H^2(X_t, \mathbb{Z})$ and $\alpha \in H^2(X_t, \mathbb{R})$ is in the Kähler cone of X_t.

The space $\mathcal{M}_X$ is naturally endowed with a local product structure, since the parameter ω varies locally in an open set of $H^2(X_t, \mathbb{C})$ (Lemma 1.1) and $H^2(X_t, \mathbb{C})$ is a locally constant vector space independent of t.

Globally, $\mathcal{M}_X$ does not necessarily have the form of a product, on the one hand because of the monodromy that may act on $H^2(X_t, \mathbb{C})$ and on the other hand because the Kähler cone may depend on t. (It is known from [**22**], however, that it is locally constant on the complement of a countable union of hypersurfaces of the moduli space of X.)

8.1. The conjecture on mirrors. *There exists a Calabi–Yau manifold with* $\dim X' = \dim X$ *and an isomorphism*

$$M : \mathcal{M}_X \cong \mathcal{M}_{X'}$$

having as its principal property preserving the local product structure by interchanging factors. Thus deforming an X with constant complexified Kähler parameter would be equivalent to deforming the complexified Kähler parameter of X' with the complex structure of X' fixed.

It would seem more prudent to assume that this mirror isomorphism exists only on a Zariski open set of the universal covering of $\mathcal{M}_X$ and $\mathcal{M}'_X$. It should also be noted that the "conjecture" cannot be true as such, simply because there exist Calabi–Yau threefolds with $h^{2,1} = 0$, and their mirrors should be manifolds with $h^{1,1} = 0$, thus non-Kähler manifolds. It remains true that this prediction of the physicists is largely confirmed, albeit not understood mathematically (see Chapters 3 and 4).

On the infinitesimal level this property of M can be translated as follows. The local product structure of $\mathcal{M}_X$ derives from the natural decomposition

(1.47) $$T_{\mathcal{M}_X,(X,\omega)} = H^1(T_X) \oplus H^1(\Omega_X),$$

where the first term describes the infinitesimal deformations of the complex structure (Proposition 1.10) and the second term describes those of the complexified Kähler parameter. Then the differential M_* should split into a direct sum of two isomorphisms:

(1.48) $$H^1(T_X) \cong H^1(\Omega_{X'}), \quad H^1(\Omega_X) \cong H^1(T_{X'}).$$

More generally we should have a series of isomorphisms

(1.49) $$H^p\bigl(\overset{q}{\wedge} T_X\bigr) \cong H^p\bigl(\overset{q}{\wedge} \Omega_{X'}\bigr).$$

Since we have $H^p(\wedge^q T_X) \cong H^p(\wedge^{n-q} \Omega_X)$ by the triviality of K_X, we can deduce that all the Hodge numbers of X', and hence also the Betti numbers, are determined by those of X.

8.2. An example. The construction that follows was proposed independently by Borcea [5] and Voisin [20]. (Other authors [17], [19] have also worked in a similar area, studying mirror symmetry for hyper-Kähler manifolds (see 2), which will not be discussed in the present text, since we are considering only Calabi–Yau manifolds X with $h^2(\mathcal{O}_X) = 0$.)

The K3 surfaces (see [24]) are the simply connected Kähler surfaces with trivial canonical bundle. They form one of the rare classes of manifolds for which Torelli's Theorem is known.

THEOREM 1.23 (see [24]). *Let S and S' be two K3 surfaces such that there exists an isomorphism of Hodge structures $\phi : (H^2(S, \mathbb{Z}), \langle, \rangle) \cong (H^2(S', \mathbb{Z}), \langle, \rangle)$. Then S is isomorphic to S'.*

Another remarkable fact is the surjectivity of the marked period map $\mathcal{P}$, which associates with S the line

$$\langle \omega_S \rangle = H^{2,0}(S) \subset H^2(S, \mathbb{C}) \cong L_\mathbb{C},$$

where $L_\mathbb{C} = L_\mathbb{Z} \otimes \mathbb{C}$ and $L_\mathbb{Z}$ is an even unimodular lattice of rank 22 and signature $(3, 19)$. The class ω_S is subject to the polarization conditions $\langle \omega_S, \omega_S \rangle = 0$ and $\langle \omega_S, \bar{\omega}_S \rangle > 0$ (see 1.5.3), defining the domain of periods $\mathcal{D}$ constructed on $(L, \langle, \rangle)$. We have

THEOREM 1.24 (see [24]). *The marked period map $\mathcal{P}$ is onto.*

We consider families of surfaces S of type K3 obtained as a double covering of a rational surface T branched along a curve C. By the period map and the fine version of Torelli's theorem, they are essentially characterized by the existence of an involution i, acting as an automorphism of the Hodge structure on $H^2(S, \mathbb{Z})$, so that

$$i^*(\omega_S) = -\omega_S.$$

This involution should then satisfy the condition that the signature of $\langle, \rangle_{|H^2(S, \mathbb{Z})}$ be equal to $(2, \operatorname{rank} H^2(S, \mathbb{Z})^- - 2)$, where $H^2(S, \mathbb{Z})^-$ is the set of classes that are anti-invariant under i. The involution i over $(L_\mathbb{Z}, \langle, \rangle)$ being fixed, the period domain $\mathcal{D}_i$ for such a family is then equal to $\mathcal{D} \cap \mathbb{P}(L_\mathbb{C}^-)$. The family of K3 surfaces so obtained depends only on the conjugacy class of i under $\operatorname{Aut}(L_\mathbb{Z}, \langle, \rangle)$.

With such a surface S and a choice of an elliptic curve E obtained as the double covering of $\mathbb{P}^1$ branched in four points and hence endowed with an involution j over $\mathbb{P}^1$, we associate the Calabi–Yau threefold satisfying $h^2(\mathcal{O}_X) = 0$ obtained by the desingularization of the quotient $E \times S/(j, i)$, as in 7.2. We denote by N the number of components of the branch curves of $S \to T$ and by N' the sum of the genera of these components. We have

PROPOSITION 1.25 (see [20]). *The Hodge numbers of X are given by the following formula*

(1.50) $\qquad h^{1,1}(X) = 11 + 5N - N', \quad H^{2,1}(X) = 11 + 5N' - N.$

8. MIRRORS

The mirror family is thus obtained by constructing a "mirror involution" i' on $L_\mathbb{Z}$ as follows. We show that, with only a few exceptions, there exists a hyperbolic plane $P \subset L_\mathbb{Z}^-$, unique up to an automorphism of $(L_\mathbb{Z}, \langle\,,\,\rangle, i)$ according to the papers of Nikulin [14]. The involution i' is then defined by

$$i' = i \circ r_P,$$

where r_P is reflection relative to the plane P. Its conjugacy class depends only on that of i.

Still following [14], we note that the conjugacy class of i is essentially characterized by the rank of $L_\mathbb{Z}^-$ and the rank of the kernel of the reduction of $\langle\,,\,\rangle_{|L_\mathbb{Z}^-}$ modulo 2. (In fact there is an invariant of a more subtle "parity" type assuming the values 0 or 1.) These invariants can easily be calculated using the invariants N and N' of the family of K3 surfaces with involution determined by i, thus enabling us to deduce the following theorem.

THEOREM 1.26 (see [20]). *We have $N(i) = N'(i')$ and $N(i') = N'(i)$ so that by formula (1.50), for the families of Calabi–Yau manifolds X (resp. X') determined by the involution i (resp. i'), we have the inversion of Hodge numbers predicted by mirror symmetry:*

$$h^{1,1}(X) = h^{2,1}(X'), \quad h^{2,1}(X) = h^{1,1}(X').$$

One can finally associate with i a second domain $\mathcal{D}'_i$ parameterizing the "complexified i-invariant Kähler parameters" by the formula

$$\mathcal{D}'_i = \left\{ \eta \in L_\mathbb{C}^+, \langle \mathrm{Re}\,\eta, \mathrm{Re}\,\eta \rangle > 0 \right\}.$$

The author [20] has constructed mirror isomorphisms

$$M : \mathcal{D}_i \cong \mathcal{D}'_{i'}, \quad M' : \mathcal{D}'_i \cong \mathcal{D}_{i'}$$

that, by Torelli's theorem 1.23 and combined with the mirror map for the elliptic curves described in the introduction, make it possible to construct the mirror map between the subspace of $\mathcal{M}_X$ parameterizing the deformations of X of the form $\widetilde{E \times S}/(j,i)$ endowed with a complexified Kähler parameter of the form

$$\lambda = \mathrm{pr}_1^* \lambda_E + \mathrm{pr}_2^* \lambda_S,$$

where $\lambda_S \in \mathcal{D}'_i$, and the analogous space for the family $\{X'\}$ associated with i'. To such a couple (X, λ) there corresponds the period τ_E of E, the line $\langle \omega_S \rangle \in \mathcal{D}_i$, the Kähler parameter λ_E on E, and the parameter $\lambda_S \in \mathcal{D}'_i$.

As in the introduction, we construct the mirror $(E', \lambda_{E'})$ of (E, λ) by the formula

$$\tau_{E'} = i\lambda_E, \quad \lambda_{E'} = -i\tau_E$$

and on the other hand we set

$$\langle \omega_{S'} \rangle = M'(\lambda_S), \quad \lambda_{S'} = M(\langle \omega_S \rangle),$$

which provides the mirror couple (X', λ').

CHAPTER 2

"Physical" origin of the conjecture

The purpose of this chapter is to describe the ideas from mathematical physics that made it possible to exhibit the phenomenon of mirror symmetry.

We begin by describing on the classical level the field theory considered by physicists using an introduction to supervariables, which leads to the essential point: the classical invariance of action by "$N = 2$-supersymmetry" when the image Riemannian manifold is a Kähler manifold.

We then undertake to describe the thought processes that led physicists to associate $N = 2$-superconformal field theory with (X, ω) by "quantification" and the way in which mirror symmetry is constructed formally as an involution on the $N = 2$-superalgebra of infinitesimal symmetries of the action on its representations.

Finally, following Lerche–Vafa–Warner and Witten, we explain how the Dolbeault cohomology of X can be calculated starting from $N = 2$-superconformal field theory associated with (X, ω) and how the predictions concerning the comparison of the Hodge numbers and Yukawa couplings of (X, ω) and its mirror (X', ω') can be derived from this formal construction of mirror symmetry on the level of $N = 2$-superconformal theories.

1. The $N = 2$-supersymmetric σ-model

1.1. The bosonic σ-model in dimension 2. Let (M, g) be a Riemannian manifold. The bosonic σ-model in dimension 2 studies the space of maps $\phi : \Sigma \to M$, where Σ is a Riemann surface with metric γ, endowed with the (energy) action

$$(2.1) \qquad S(\phi, \gamma) = \int_\Sigma |d\phi|^2.$$

This action is invariant under *change of scale*, that is,

$$S(\phi, \gamma) = S(\phi, e^h \cdot \gamma).$$

for every real-valued function h on Σ.

In addition S is invariant under the diffeomorphisms of Σ:

$$S(\phi, \gamma) = S(\phi \circ \psi, \psi^* \gamma).$$

Thus S is invariant under *conformal transformations*

$$(2.2) \qquad S(\phi, \gamma) = S(\phi \circ \psi, \gamma)$$

for a conformal transformation ψ of Σ.

When (M, g) is a Kähler manifold, $S(\phi)$ can be rewritten as follows. The complex structures on M and Σ make it possible to define $\bar{\partial}\phi \in \Omega^{0,1}_\Sigma \otimes \phi^* T^{1,0}_M$ as the antilinear part of $d\phi$. If α is the Kähler form of the metric on M, we then have

$$(2.3) \qquad S(\phi) = \int_\Sigma \phi^* \alpha + \int_\Sigma 2|\bar{\partial}\phi|^2,$$

where the first term is constant on deformations of ϕ such that $\phi_{\partial\Sigma}$ remains constant, since α is closed. If a closed 2-form β is also given on M, one can also consider the action

$$S(\phi) = \int_\Sigma |d\phi|^2 + \int_\Sigma \phi^*(\beta). \tag{2.4}$$

The second term is constant under deformations of ϕ with constant boundary, and hence does not enter into the Euler–Lagrange equations that describe the critical points of S; in addition, if β is modified by the addition of an exact form, the action is modified only by an integral over the boundary of Σ. Finally, if (M,g) is a Kähler manifold with Kähler form α, a closed 2-form β provides a complexified Kähler parameter (see 1.8) $\omega = \alpha + i\beta$, and, when (2.3) and (2.4) are combined, an action

$$S(\phi) = \int_\Sigma \phi^*\omega + \int_\Sigma 2|\bar\partial\phi|^2, \tag{2.5}$$

where the norm in the second term is computed using the Hermitian metrics induced by γ on Σ (in fact the integral depends only on the conformal class of γ) and g_α on M.

1.2. Supervariables. Let Λ be the exterior algebra of infinite dimension,

$$\Lambda = \varinjlim_k \wedge(\mathbb{R}^k). \tag{2.6}$$

One can write

$$\Lambda = \Lambda^+ \oplus \Lambda^-$$

(decomposition with respect to the parity of the degree) and the commutative algebra Λ^+ is naturally a direct sum

$$\Lambda^+ = \mathbb{R} \oplus \Lambda_s^+, \tag{2.7}$$

where $\mathbb{R} = \Lambda^0$ and $\Lambda_s^+ = \Lambda_{>0}^+$ consists of nilpotent elements. We set

$$\mathbb{R}^{m,n} = (\Lambda^+)^m \times (\Lambda^-)^n. \tag{2.8}$$

This is the prototype of *supermanifolds of dimension* (m,n).

Its topology is the following. The direct-sum (2.7) provides a projection $\pi : \mathbb{R}^{m,n} \to \mathbb{R}^m$, and the open sets in $\mathbb{R}^{m,n}$ are the $\pi^{-1}(U)$ with $U \subset \mathbb{R}^m$ open.

The *superdifferentiable* functions on an open set U of $\mathbb{R}^{m,n}$ are described as follows. We use the notation

- $x_i = x_i^0 + x_i^s$, $i = 1,\ldots,m$, for the even coordinates obtained by projecting $\mathbb{R}^{m,n}$ on its components Λ^+ and
- θ^j, $j = 1,\ldots,n$, for the odd coordinates.

A superdifferentiable function Φ (with values in Λ) on U should then be written as a series in the θ^j (necessarily finite since the θ^j anticommute):

$$\Phi = \sum_{I=\{i_1<\cdots<i_k\}} \Phi_I(x_1,\ldots,x_m)\theta^{i_1}\cdots\theta^{i_k}, \tag{2.9}$$

where Φ_I has values in Λ, depends only on the even coordinates, and is superdifferentiable in the following sense. Let ϕ_I be the restriction of Φ_I to $\{x_i^s = 0;\ i =$

$1, \ldots, m\}$. Then ϕ_I should be of class $\mathcal{C}^\infty$, and Φ_I is the Taylor expansion of ϕ_I in the variables x_i^s:

$$(2.10) \qquad \Phi_I(x_1, \ldots, x_m) = \sum_J \frac{1}{J!} \frac{\partial^J \phi_I}{\partial x^J}(x_1^0, \ldots, x_m^0)(x^s)^J,$$

where $J = (j_1, \ldots, j_m)$ and $(x^s)^J = (x_1^s)^{j_1} \cdots (x_m^s)^{j_m}$.

A supermanifold of dimension (m, n) should be covered by open sets homeomorphic to open sets in $\mathbb{R}^{m,n}$, and the transition functions should be given by invertible superdifferentiable transformations:

$$(2.11) \qquad \begin{cases} \theta'^j = \Psi_j(x_1, \ldots, x_m, \theta^1, \ldots, \theta^n), \\ x_i' = \Phi_i(x_1, \ldots, x_m, \theta^1, \ldots, \theta^n), \end{cases}$$

where the Ψ_j are superdifferentiable and have values in Λ^- and the Φ_i are superdifferentiable and have values in Λ^+. The invertibility of this transformation is guaranteed locally by the non-vanishing of the "superdeterminant" (see [40]) of the Jacobian matrix.

Given an ordinary manifold M of class $\mathcal{C}^\infty$, one can construct a supermanifold of dimension $(m, 0)$ with $m = \dim M$, by replacing the open sets of M homeomorphic to open sets U of $\mathbb{R}^m$ by $\pi^{-1}(U)$ and taking as transition functions between two such open sets the Taylor expansions (2.10) of the transition functions of the corresponding open sets in M.

DEFINITION 2.1. The *superplane* is the supermanifold $\mathbb{R}^{2,2}$.

It admits global coordinates x_i, θ^α for $i = 1, 2$ and $\alpha = 1, 2$. Indeed, $\mathbb{R}^2$ being endowed with the Euclidean metric, the coordinates θ^α must be understood as corresponding to the choice of a base e_α of the space $\mathcal{S}$ consisting of spinors of $\mathbb{R}^2$.

It will be recalled (see [34]) that the spinor space of $(\mathbb{R}^n, \langle, \rangle)$ for n even is the unique irreducible representation of the Clifford algebra $C(n)$ generated by $\mathbb{R}^n$,

$$C(n) = \bigoplus_r (\mathbb{R}^n)^{\otimes r} / I,$$

where I is generated by the relations

$$x \cdot x = -\langle x, x, \rangle 1, \quad x \in \mathbb{R}^n.$$

The vectors $u \in \mathbb{R}^n$ with n even act on the spinor space $\mathcal{S}$ (the Clifford action) because $\mathbb{R}^n$ is included in the Clifford algebra $C(n)$.

The couple (θ^1, θ^2) thus describes a spinor of $\mathbb{R}^2$ with coefficients in Λ^-. In this way, all the considerations that follow below have a meaning on any Riemann surface endowed with a spinor structure that determines a spinor bundle $\mathcal{S}$ on which the tangent vectors act by the Clifford action.

1.3. Supersymmetry.

DEFINITION 2.2. A *supersymmetry generator* is an odd vector field.

This has the following meaning: the algebra of superdifferentiable functions on a supermanifold is a superalgebra

$$\mathcal{F} = \mathcal{F}^+ \oplus \mathcal{F}^-,$$

where $\mathcal{F}^+$ (resp. $\mathcal{F}^-$) is the set of functions with values in Λ^+ (resp. Λ^-).

This means that the elements of $\mathcal{F}^+$, called *even*, commute with all the elements of $\mathcal{F}$, while the elements of $\mathcal{F}^-$, called *odd*, anticommute. The decomposition $\mathcal{F} = \mathcal{F}^+ \oplus \mathcal{F}^-$ is compatible with the algebra structure, so that $\mathcal{F}$ is $\mathbb{Z}/2\mathbb{Z}$-graded. An odd vector field X is an odd derivation of $\mathcal{F}$, which means that it satisfies the *twisted Leibniz rule*:

(2.12) $$X(fg) = (Xf)g + (-1)^{\deg f} f(Xg)$$

for homogeneous elements f, g of $\mathcal{F}$.

It can be verified that if X and Y are odd derivations, the supercommutator

$$\{X, Y\} := X \circ Y + Y \circ X$$

is an even derivation.

On $\mathbb{R}^{2,2}$ we consider the "spinor derivations" (odd vector fields whose covariant character is spinor):

(2.13) $$D_\alpha = \frac{\partial}{\partial \theta^\alpha} + i \sum_\beta \theta^\beta \frac{\partial}{\partial_{\alpha\beta}},$$

where $\partial/\partial_{\alpha\beta}$ is the constant vector field on $\mathbb{R}^2$ obtained by *contraction* of the spinors e_α, e_β, the contraction being the composition

(2.14) $$\mathcal{S} \otimes \mathcal{S} \cong \mathrm{Hom}(\mathcal{S}, \mathcal{S}) \longrightarrow T^*_{\mathbb{R}^2} \cong T_{\mathbb{R}^2},$$

where the two isomorphisms are given by the metrics, and the arrow in the middle is the dual of the map given by the Clifford action.

The following supercommutation rule can be verified immediately:

(2.15) $$\{D_\alpha, D_\beta\} = 2i \frac{\partial}{\partial_{\alpha\beta}}.$$

DEFINITION 2.3. The *Poincaré $N = 1$-superalgebra* is the Lie superalgebra generated as a vector space by the D_α and by the Lie algebra of the group of translations and rotations of $\mathbb{R}^2$.

One can easily show that this space is stable under the (super)commutator. (For example, (2.15) shows that the supercommutator of odd generators is in the usual Poincaré algebra.) Because the D_α are odd fields of spinor type, they do not act on the scalar "superfields" (that is, the even superdifferentiable functions).

In contrast, if (ϵ_α) is a section of the dual spinor bundle with coefficients in Λ^-, one can construct an infinitesimal action of

$$Q = \sum_\alpha \epsilon_\alpha D_\alpha,$$

called a *supersymmetric transformation*, on scalar superfields by the rule:

(2.16) $$\delta_Q \Phi = \sum_\alpha \epsilon_\alpha D_\alpha \Phi.$$

Let Φ be a scalar superfield on $\mathbb{R}^{2,2}$. By (2.9) this superfield admits the following expression

(2.17) $$\Phi = \phi + \sum_\alpha \theta^\alpha \psi_\alpha + \theta^\alpha \theta^\beta F_{\alpha\beta},$$

where ϕ is an even function called the *bosonic component* of Φ, the ψ_α are odd (of spinor type) and called the *fermionic components* of Φ; and $F_{\alpha\beta} = F_{\beta\alpha}$, whose components are called the *auxiliary parameters*, is even and should be regarded as a section of $\wedge^2(\mathcal{S})$. By definition of D_α we have

(2.18) $$\delta_Q \Phi = \sum_\alpha \epsilon_\alpha \psi_\alpha + \sum_{\alpha,\beta} \epsilon_\alpha \theta^\beta F_{\alpha\beta} + i \sum_{\alpha,\beta} \epsilon_\alpha \theta^\beta \partial_{\alpha\beta} \phi$$
+terms of degree 2 in θ.

which gives immediately the following formula for the action of δ_Q on the fermionic and bosonic components of Φ:

(2.19) $$\delta_Q \phi = \sum_\alpha \epsilon_\alpha \psi_\alpha, \quad \delta_Q \psi_\alpha = \sum_\beta \epsilon_\beta (F_{\alpha\beta} - i\partial_{\alpha\beta}\phi).$$

1.4. The Berezin integral. The Berezin integral of an even superdifferentiable function Φ of compact support on $\mathbb{R}^{m,n}$ consists of the integral of the coefficient of the highest-degree term in θ in the expansion (2.9) of Φ, taken in the usual sense on $\mathbb{R}^m$. If the function Φ is expanded as in (2.17), we thus have on $\mathbb{R}^{2,2}$

(2.20) $$\int_{\mathbb{R}^{2,2}} \Phi\, d^2x\, d^2\theta = \int_{\mathbb{R}^{2,2}} F_{\alpha\beta}\, d^2x,$$

where $F_{\alpha\beta} \in \wedge^2(\mathcal{S})$ can be regarded as a function because of the natural trivialization of the bundle $\wedge^2(\mathcal{S})$. We then have the following lemma:

LEMMA 2.4. *Let Φ be a scalar superfield with compact support and (ϵ_α) a section of $\mathcal{S}^*$ (with coefficients in Λ^-) annihilated by the Dirac operator (that is, (ϵ_α) is the sum of a holomorphic spinor and an anti-holomorphic spinor [34]). Then $\int_{\mathbb{R}^{2,2}} \delta_Q \Phi = 0$.*

This follows from the fact that the infinitesimal variation $\delta_Q F_{\alpha\beta}$ is then a divergence on $\mathbb{R}^2$.

This lemma shows that the Berezin integral is invariant under supersymmetric transformations having holomorphic or antiholomorphic parameters, and the latter is the reason why the supersymmetric σ-model is invariant under supersymmetry.

1.5. The supersymmetric σ-model. We consider a Riemannian manifold (M, g) of class $\mathcal{C}^\infty$, which one can regard as an even supermanifold (see 2.1.2). The scalar superfields with values in M (that is, superdifferentiable maps $\Phi : \mathbb{R}^{2,2} \to M$, where $\mathbb{R}^{2,2}$ can also be replaced by any Riemann supersurface) are endowed with the following action:

(2.21) $$S(\Phi) = \int_{\mathbb{R}^{2,2}} \sum_{i,j,\alpha} g_{ij} \circ \Phi D^\alpha \Phi^i D_\alpha \Phi^j\, d^2x\, d^2\theta,$$

where the indices i and j refer to a choice of coordinates y^i on M (the expression to be integrated being in fact independent of the coordinates), and $D^\alpha = \sum_\beta C^{\alpha\beta} D_\beta$, where $C_{\alpha\beta}$ describes the metric on the spinor bundle $\mathcal{S}$ so that the term to be integrated is actually an inner product

$$\langle (D_\alpha \Phi^i), (D_\alpha \Phi^i) \rangle,$$

where $(D_\alpha \Phi^i) \in \mathcal{S}^* \otimes \phi^*(T_M)$. It should be noted here that $g_{ij} \circ \Phi$ is obtained starting from $g_{ij} \circ \phi$ (where ϕ is the bosonic component of Φ) by a Taylor expansions in the part of degree higher than 1 in the thetas, as in (2.10).

The bosonic part of $S(\Phi)$, that is, the part independent of θ, is precisely the action $S(\phi)$ of (2.1). This action possesses both conformal invariance (2.2) and invariance under supersymmetric transformations having holomorphic or antiholomorphic parameter, as follows from Lemma 2.4.

The superfield Φ admits an expansion in components (see (2.17))

$$\Phi^j = \phi^j + \sum_\alpha \theta^\alpha \psi^j_\alpha + \theta^\alpha \theta^\beta F^j_{\alpha\beta}, \tag{2.22}$$

where, by definition of a superdifferentiable function, the m-tuple (ψ^j_α) transforms like a section of $\phi^* TM^{\mathbb{C}} \otimes \mathcal{S}$.

The "auxiliary parameters" $F^j_{\alpha\beta}$ can be eliminated algebraically by applying the field equations. Indeed, it is easy to verify, by expanding $S(\Phi)$ and writing it in the form $S(\Phi) = \int_{\mathbb{R}^2} L(\Phi)$, that $L(\Phi)$ is an inhomogeneous polynomial of degree 2 in the $F^j_{\alpha\beta}$, the homogeneous term of degree 2 being equal to

$$\sum_{ij\alpha\beta} C^{\alpha\beta} g_{ij}(\phi) F^i_{\alpha\beta} F^j_{\alpha\beta},$$

so that one can obtain an action $S(\phi, \psi)$ that now depends only on the bosonic and fermionic components of Φ by finding the exremal value of $S(\Phi)$ with respect to $F^j_{\alpha\beta}$. The result is the following formula:

$$F^j_{\alpha\beta} = \sum_{ik} \Gamma^j_{ik} \psi^i_\alpha \psi^k_\beta, \tag{2.23}$$

where the Γ^j_{ik} are the Christoffel symbols of g in the coordinates y_j.

Adopting the holomorphic notation

$$z = x_1 + ix_2, \quad \partial_z = \frac{1}{2}(\partial_{x_1} - i\partial_{x_2})$$

and the decomposition of complexified spinors into holomorphic $(dz^{1/2})$ and antiholomorphic $(d\bar{z}^{1/2})$, we find that the sections (ψ^j_α) of $\phi^* TM^{\mathbb{C}} \otimes \mathcal{S}$ will be decomposed into

$$\psi^i_+ \in \phi^* TM^{\mathbb{C}} \otimes K^{1/2} \quad \text{and} \quad \psi^i_- \in \phi^* TM^{\mathbb{C}} \otimes \overline{K}^{1/2},$$

where K is the canonical bundle of $\mathbb{R}^2 \cong \mathbb{C}$, or, more generally, of the Riemann surface on which one is working. We then have the following formula:

$$S(\phi, \psi) = \int_{\mathbb{R}^2} 4\big(g_{kl}(\phi)\partial_z \phi^k \partial_{\bar{z}} \phi^l + \sqrt{-1} g_{kl}(\phi, \psi^k_- \nabla_z \psi^l_- \tag{2.24}$$

$$+ \sqrt{-1} g_{kl}(\phi) \psi^k_- \nabla_{\bar{z}} \psi^l_+ + \frac{1}{2} R_{klmn} \psi^k_+ \psi^l_+ \psi^m_- \psi^n_-\big)\, dx_1\, dx_2.$$

The appearance of the Levi–Civita connection of M and its curvature tensor R_{ijkl} in this formula is due to the Taylor-series expansion of $g_{ij}(\Phi)$, which causes the derivatives of the metric up to second order to appear. Since the action $S(\phi, \psi)$ is obtained by finding the extremum with respect to $F^j_{\alpha\beta}$, it remains invariant under the supersymmetric transformations (2.19) when $F^j_{\alpha\beta}$ is replaced by its value (2.23). In the holomorphic notation (2.19) then becomes:

$$\begin{cases} \delta_Q \phi^k = \epsilon \psi^k_+ + \bar{\epsilon} \psi^k_-, \\ \delta_Q \psi^k_+ = i\epsilon \partial_z \phi^k - \bar{\epsilon} \sum_{lm} \psi^l_- \Gamma^k_{lm} \psi^m_+, \\ \delta_Q \psi^k_- = i\bar{\epsilon} \partial_{\bar{z}} \phi^k - \epsilon \sum_{lm} \psi^l_+ \Gamma^k_{lm} \psi^m_-, \end{cases} \tag{2.25}$$

where ϵ (resp. $\bar\epsilon$) is a holomorphic section of $K^{-1/2}$ (resp. antiholomorphic section of $\overline{K}^{-1/2}$).

1.6. $N = 2$-supersymmetry. Alvarez-Gaumé and Freedman [26] have shown that the supersymmetric σ-model possesses a second type of invariance under supersymmetry when (M,g) is a Kähler manifold. The $N = 2$-supersymmetric Poincaré algebra is the Lie subalgebra obtained by adjoining a second set of odd generators D'_α satisfying the rule (2.15) to the $N = 1$-supersymmetric Poincaré algebra (see Definition 2.3):

$$\{D'_\alpha, D'_\beta\} = 2i\partial_{\alpha\beta}, \tag{2.26}$$

generators that "supercommute" with the D_α:

$$\{D'_\alpha, D_\beta\} = 0. \tag{2.27}$$

Alvarez-Gaumé and Freedman attempt to discover the conditions under which the action $S(\phi,\psi)$ is invariant with respect to supersymmetric transformations of the form

$$\begin{cases} \delta'_Q \phi^i = \sum_j f^i_j(\epsilon\psi^j_+ + \bar\epsilon\psi^j_-), \\ \delta'_Q \psi^i_+ = i\epsilon\sum_j h^i_j(\partial_z\phi^j) - \bar\epsilon\sum_{kl}\psi^k_- S^i_{kl}\psi^l_+, \\ \delta'_Q \psi^i_- = i\bar\epsilon\sum_j h^i_j(\partial_{\bar z}\phi^j) - \epsilon\sum_{kl}\psi^k_+ S^i_{kl}\psi^l_-, \end{cases} \tag{2.28}$$

where f^i_j and h^i_j are assumed to be sections of $\mathrm{End}\,TM$ such that the components δ_α, δ'_α of δ_Q, δ'_Q (where $\delta_Q = \sum_\alpha \epsilon_\alpha \delta_\alpha$ and $\delta'_Q = \sum_\alpha \epsilon_\alpha \delta'_\alpha$ if we use nonholomorphic notation) satisfy the anticommutation rules (2.26) and (2.27).

But it is easy to see that the condition (2.26) implies that $f^i_j = (h^i_j)^{-1}$, while the rule (2.27) implies that $f^i_j = -h^i_j$, so that f^i_j should define a pseudocomplex structure J on M. Finally, the condition that $S(\phi,\psi)$ be invariant under (2.28) implies by an immediate computation that J is parallel for ∇ and that the metric g_{ij} is J-invariant (and hence that the metric is a Kähler metric by Lemma 1.2). It can then be verified that for $S^i_{jk} = \sum_l \Gamma^i_{jl} f^l_k$ we do indeed have the required invariance.

2. Quantification

2.1. The symplectic structure on the space of classical solutions. We prescribe a theory of fields $\phi : V \to M$, $\dim V = d$ with an action

$$S(\phi) = \int_{V^d} L\bigl(x, \phi(x), \partial_i\phi, \dots\bigr)\,d\mu \tag{2.29}$$

where μ is a volume element on V and L is a Lagrangian defined on the space of n-jets of maps of V into M. (In what follows we shall assume that $n = 1$, which is the case of most interest to us.)

We consider the space $\mathfrak{M}$ of solutions of the Euler–Lagrange equations, that is, the space of critical points of S. Taking coordinates y_ν on M and x_i on V, we have coordinates x_i, y_ν, y^i_ν on the space of 1-jets from V into M, and ϕ is a critical point of S if and only if for every differentiable section $X = (X_\nu)$ of $\phi^*(T_m)$ that vanishes on ∂V the following relation holds:

$$\delta_X S := \int_V \Big(\sum_\nu X_\nu \frac{\partial L}{\partial y_\nu} + \sum_{i,\nu} \frac{\partial X_\nu}{\partial x_i} \frac{\partial L}{\partial y_i} \Big) \, d\mu = 0.$$

Setting $d\mu_i = (-1)^i \text{int}\,(\partial/\partial x_i)\, d\mu$, we find, on integrating by parts, the formula

$$\delta_X S = \int_{\partial V} \sum_\nu X_\nu \frac{\partial L}{\partial y_\nu^i} \, d\mu_i + \int_V \sum_\nu X_\nu \Big(\frac{\partial L}{\partial y_\nu} - \sum_i \frac{\partial^2 L}{\partial x_i \partial y_\nu^i} \Big) \, d\mu,$$

which holds for every increment $X \in \phi^* T_M$ of ϕ, so that the vanishing of $\delta_X S$ when $X_{|\partial V} = 0$ is equivalent to the equations

$$\forall \nu, \quad \frac{\partial L}{\partial y_\nu} - \sum_i \frac{\partial^2 L}{\partial x_i \partial y_\nu^i} = 0.$$

When these equations, called the *Euler–Lagrange* equations, are satisfied, we have for every X in $\mathcal{C}^\infty(\phi^* T_M)$ a form η_X of degree $(d-1)$ on V,

$$\eta_X = \sum_\nu X_\nu \frac{\partial L}{\partial y_\nu^i} \, d\mu_i,$$

such that

$$\forall V' \subset V, \quad \delta_{X_{|V'}} S(\phi_{|V'}) = \int_{\partial V'} \eta_X.$$

It can be verified that η_X is independent of the choice of coordinates, and is therefore globally defined. If we have prescribed a homology class of hypersurfaces $W \subset V$ (one is often interested in the case when one time coordinate on V is distinguished, $V \cong W \times [a, b]$, so that the Euler–Lagrange equations become the evolution equations for $\phi_{|W}$ and assigning W forms part of the problem), we can then construct a closed 2-form ω on $\mathcal{M}$ (which induces a symplectic structure on an adequate quotient of $\mathcal{M}$) as follows. Having chosen a hypersurface W in this homology class, we define a 1-form Θ_W on $\mathcal{M}$ by the formula

$$\Theta_W(X) = \int_W \eta_X.$$

If W is changed to W' with $\partial V_{W',W} = W' - W$, we have

$$\Theta_{W'}(X) - \Theta_W(X) = \int_{\partial V_{W',W}} \eta_X = \delta_X S(\phi_{|V_{W',W}}),$$

so that the difference $\Theta'_W - \Theta_W$ is an exact form on $\mathcal{M}$, which implies that $\omega = d\Theta_W$ is independent of the choice of W.

This symplectic structure on $\mathcal{M}$ essentially explains Noether's theorem, which associates a function on $\mathcal{M}$ with every infinitesimal symmetry of S in a manner compatible with the Lie bracket and the Poisson bracket. (If $V \cong W \times [a, b]$, this function becomes a first integral of the evolution equations.) In fact, we have only to consider the function

$$h_X(\phi) = \int_W \eta_X.$$

As X is an infinitesimal symmetry of S, we have $\delta_{X_{|V'}} S(\phi_{|V'}) = 0$ for all $V' \subset V$, and hence η_X is closed, which implies that h_X is independent of the choice of W.

By quantification one attempts to represent a certain set of functionals known as "observables" endowed with the Poisson bracket defined by ω on a Hilbert space $\mathcal{H}$ of "states." (In general, only a projective representation is meant.) In addition it would also be desirable to realize the "correlation functions" $\langle f_1(x_1) \cdots f_k(x_k) \rangle$, computed by definition using the inner product on $\mathcal{H}$ as functional integrals

$$(2.30) \qquad \int_\phi f_1(\phi(x_1)) \ldots f_k(\phi(x_k)) e^{-S(\phi)} \, d\phi,$$

where the f_i are certain functions on M and $x_i \in V$.

More precisely, supposing for the sake of simplicity that $d = 2$, we should have a state $\Omega \in \mathcal{H}$ called the "empty" state, such that for functions f_i on M and $x_i \in \mathbb{P}^1$ we would have

$$(2.31) \qquad \begin{aligned} \langle f_1(x_1) \cdots f_k(x_k) \rangle &= \int_{\phi: \mathbb{P}^1 \to M} f_1(\phi(x_1)) \cdots f_k(\phi(x_k)) e^{-S(\phi)} \, d\phi \\ &= \langle \Phi_{f_1(x_1)} \circ \cdots \circ \Phi_{f_k(x_k)}(\Omega), \Omega \rangle_\mathcal{H}, \end{aligned}$$

where $\Phi_{f_i(x_i)}$ is the operator associated with the functional $\phi \mapsto f_i(\phi(x_i))$.

In the case of the bosonic σ-model in dimension 2 we find by (2.2) that the action is invariant under conformal transformations, and thus if we take the punctured disk Δ^* of $\mathbb{C}^*$ as Σ, we have an infinite number of infinitesimal symmetries of the action, corresponding to the Lie algebra known as the *Virasoro* algebra, of holomorphic or antiholomorphic vector fields on Δ^* generated by the $z^n \partial/\partial z$ and the $\bar{z}^n \partial/\partial \bar{z}$ for $n \in \mathbb{Z}$. The conformal invariance is preserved at the quantum level if the corresponding Lie algebra can be represented on $\mathcal{H}$ in such a way that its action on observables corresponds to the operator bracket on $\mathcal{H}$.

2.2. Conformal field theory. If the Virasoro algebra (or a central extension) can be represented, one hopes by "integration" to construct a conformal field theory, that is, the following system of assignments, formalized by Segal (see [**29**], [**36**], and [**37**]).

If Σ is a Riemann surface with metric γ and boundaries l_i for $i \in I$ and l_j for $j \in J$, endowed with a parameterization $\chi_i^+ : S^1 \cong l_i$ preserving the orientation for $i \in I$ and $\chi_j^- : S^1 \cong l_j$, reversing the orientation for $j \in J$, γ being normalized near the boundaries of Σ, one should associate an "amplitude"

$$(2.32) \qquad A(\Sigma, \gamma, \chi_i^+, \chi_j^-) \in \mathrm{Hom}\left(\mathcal{H}^{\otimes I}, \mathcal{H}^{\otimes J}\right),$$

where $\mathcal{H}$ is a Hermitian space. One should regard these amplitudes has having been obtained by the functional integral in the following way: with $(\Sigma, \gamma, \chi_i^+, \chi_j^-)$ the integral

$$(2.33) \qquad \int_{\substack{\phi: \Sigma \to M \\ \phi \circ \chi_l = \eta_l}} \exp\left(-S(\phi)\right) d\phi$$

associates a function $\Psi_{\Sigma, \gamma, \chi_i^+, \chi_j^-}$ on the product $LM^I \times LM^J$ that parameterizes the

$$\eta_i : S^1 \to M, \quad (i \in I), \quad \eta_j : S^1 \to M, \quad (j \in J).$$

Assuming that we now had a suitable Hermitian space $\mathcal{H}$ of functions on LM, then $\Psi_{\Sigma, \gamma, \chi_i^+, \chi_j^-}$ would the kernel that makes it possible to construct $A(\Sigma, \gamma, \chi_i^+, \chi_j^-)$.

These amplitudes should satisfy the following axioms.

(i) Invariance under the diffeomorphisms of $(\Sigma, \gamma, \chi_i^+, \chi_j^-)$.

(ii) Gluing: let $(\Sigma, \gamma, \chi_i^+, \chi_j^-)$ and $(\Sigma', \gamma', \chi_{i'}'^+, \chi_{j'}'^-)$ be data, as above, and let $(\Sigma'', \gamma'', \chi_{i''}''^+, \chi_{j''}''^-)$ be obtained by gluing Σ and Σ' together along the curves $l_{i_1} \subset \Sigma$, $l_{j_1'} \subset \Sigma'$ using the isomorphism $\chi_{j_1'}'^- \circ (\chi_{i_1}^+)^{-1}$. The subscripts i'' are in $I'' = I \cup I' - \{i_1\}$, and the j'' are in $J'' = J \cup J' - \{j_1'\}$. We should then have

(2.34) $\quad A(\Sigma'', \gamma'', \chi_{i''}''^+, \chi_{j''}''^-) = \text{Trace}_{i_1, j_1'} A(\Sigma', \gamma', \chi_{i'}'^+, \chi_{j'}'^-) \otimes A(\Sigma, \gamma, \chi_i^+, \chi_j^-)$

in $\text{Hom}(\mathcal{H}^{\otimes I''}, \mathcal{H}^{\otimes J''})$.

(iii) Conformal invariance: for a real-valued function h on Σ vanishing near the boundary we should have

(2.35) $\quad\quad\quad\quad A(\Sigma, e^h \gamma, \chi_i^+, \chi_j^-) = C(h) A(\Sigma, \gamma, \chi_i^+, \chi_j^-).$

By these axioms all the amplitudes should be computable using the amplitudes for the disk (which give the "vacuum"), for the sphere with three disks removed, and for annuli with parameterized boundaries, one "leaving" and the other "returning."

The fact that one can consider the representations of the Virasoro algebra as an infinitesimal version of conformal field theories is due to the possibility of considering the semi-group $\mathcal{A}$ of annuli with parameterized boundaries as a substitute for a "complexification of $\text{Diff}^+ S^1$" (so that the Virasoro algebra is the Lie algebra of this semigroup).

The holomorphic structure on this semigroup is given by describing such an annulus by two maps $f_0, f_1 : D \to \mathbb{P}^1$, which extend to smooth maps on the boundary $\partial D \cong S^1$ of the disk D and satisfy $f_0(0) = 0$, $f_1(0) = \infty$, $\text{Im}\, f_0 \cap \text{Im}\, f_1 = \emptyset$ modulo the action of $\mathbb{C}^* = \text{Aut}(\mathbb{P}^1, \{0, \infty\})$. We then set

$$A = \mathbb{P}^1 - (\text{Im}\, f_0 \cup \text{Im}\, f_1),$$

with the parameterization of the boundaries given by $f_{i|\partial D}$. The group $\text{Diff}^+ S^1$ is regarded as a boundary of that space $(\text{Im}\, f_0 \cup \text{Im}\, f_1)$ and the tangent space to $\mathcal{A}$ along this boundary can be identified with the complexification of the Lie algebra of $\text{Diff}^+ S^1$, that is, the Virasoro algebra.

Physicists give arguments to show that the construction of such a conformal theory through quantification of the bosonic σ-model in dimension 2 with values in (M, g) (recall that M is compact) is possible since the Ricci curvature of (M, g) is zero.

They likewise think that if (M, g) is Kähler manifold with Ricci curvature zero, one can obtain, by quantification of the $N = 2$-supersymmetric σ-model, a representation of a central extension of the "Virasoro $N = 2$-superalgebra," that is, the Lie superalgebra of infinitesimal symmetries of the action (2.21) containing the tangent space to the conformal transformations ((anti)-holomorphic vector fields on the punctured disk) and the two families of infinitesimal supersymmetries (parameterized by (anti)-holomorphic spinors).

One can also include a term of the form (2.4) in the action $S(\Phi)$, leading to the following conclusion.

Let X be a Calabi–Yau manifold satisfying the condition $h^2(X, \mathcal{O}_X) = 0$, and let $\omega = \alpha + i\beta$ be a complexified Kähler parameter on X (see 1.8); α determines a Kähler–Einstein metric (Theorem 1.5), and β contributes to the action as in (2.5). These data should make it possible to construct an $N = 2$-superconformal field theory having central charge $c = 3n$ with $n = \dim X$ by quantification of the

$N = 2$-supersymmetric σ-model constructed in (2.24) modified by the adjunction of the term $i \int_\Sigma \phi^*(\beta)$. (The central charge is a coefficient associated with the representation of the central extension of the Virasoro superalgebra (§2.4).) The fact that β can be chosen modulo $2\pi H^2(X, \mathbb{Z})$ results from the fact that $\exp\big({-}S(\phi, \psi)\big)$ depends on β only modulo $2\pi H^2(X, \mathbb{Z})$, at least if we are working with compact surfaces Σ.

3. Gepner's conjecture

Gepner [30] has conjectured that the correspondence between Calabi–Yau manifolds endowed with a complexified Kähler parameter and the theory of $N = 2$-superconformal fields whose construction was sketched above is bijective if we consider only superconformal theories with U(1) integer charges. (See 2.4 for the meaning of this expression.) In fact, the mirror phenomenon is precisely the correction that needs to be added to this claim.

The base of the conjecture is the following construction. Gepner begins with a certain discrete series of $N = 2$-superconformal theories E_k having central charge $c_k = 3k/(k+2)$ for $k \in \mathbb{N}$. He then constructs $N = 2$-superconformal theories $E_{(k_1,\ldots,k_r)}$ having central charges

$$c_{(k_1,\ldots,k_r)} = \sum_i c_{k_i} = 3 \sum_i \frac{k_i}{k_i + 2}$$

taking essentially an adequate subspace of the tensor product of the E_{k_i}. We then set

$$M = \operatorname{ppcm}(k_i + 2)_{i=1,\ldots,r}.$$

LEMMA 2.5. *The condition* $3 \sum_i \dfrac{k_i}{k_i + 2} = 3(r-2)$ *is equivalent to the condition that the hypersurfaces* $Y \subset \mathbb{P}(M/(k_i+2))$ *of degree M have trivial canonical bundle.*

Here $\mathbb{P}(M/(k_i+2))$ is the *inhomogeneous projective space* defined as the quotient of $\mathbb{C}^r - \{0\}$ under the action of $\mathbb{C}^*$:

$$g_z\big((x_1, \ldots, x_r)\big) = \big(z^{M/(k_1+2)} x_1, \ldots, z^{M/(k_r+2)} x_r\big).$$

If the projective coordinate X_i is of degree $M/(k_i + 2)$, the hypersurfaces of degree M are defined by the polynomials $F(X_1, \ldots, X_r)$ generated by the monomials $X_1^{i_1} \cdots X_r^{i_r}$ satisfying the condition

$$\sum_l i_l \frac{M}{k_l + 2} = M.$$

If we ignore the singularities of $\mathbb{P}(M/(k_i + 2))$, the proof of Lemma 2.5 is the following. One can compute the canonical bundle of $\mathbb{P}(M/(k_i + 2))$ (at least in codimension 2) as follows. There is a natural map

(2.36) $$\phi : \mathbb{P}^{r-1} \to \mathbb{P}(M/(k_i + 2))$$

obtained as the passage to the quotient of

(2.37) $$\begin{cases} \psi : \mathbb{C}^r - \{0\} \longrightarrow \mathbb{C}^r - \{0\}, \\ (x_1, \ldots, x_r) \mapsto \big(x_1^{M/(k_1+2)}, \ldots, x_r^{M/(k_r+2)}\big). \end{cases}$$

The map ϕ clearly ramifies with order $M/(k_i + 2) - 1$ along the hyperplane $H_i = \{X_i = 0\}$. We thus obtain

$$\phi^* K_{\mathbb{P}(M/(k_i+2))} = K_{\mathbb{P}^{r-1}} - \sum_i \left(\frac{M}{k_i + 2} - 1\right) H_i$$
(2.38)
$$= -\left(\sum_i \frac{M}{k_i + 2}\right) H.$$

A hypersurface $Y \subset \mathbb{P}(M/(k_i + 2))$ of degree d thus has by the adjunction formula a trivial canonical bundle (in codimension 2) if

$$d = \sum_i \frac{M}{k_i + 2}.$$

In particular, the hypersurfaces of degree M have trivial canonical bundle (in codimension 2) when $\sum_i (k_i + 2)^{-1} = 1$, which proves the lemma. $\square$

Now for "physical" reasons Gepner has to impose this condition on the charge so that $E_{(k_1,\ldots,k_r)}$ is obtained by quantification of the $N = 2$-supersymmetric σ-model with values in a Calabi–Yau manifold of dimension $(r - 2)$.

In fact, Gepner conjectures more precisely that $E_{(k_1,\ldots,k_r)}$ corresponds to the Fermat hypersurface $Y \subset \mathbb{P}(M/(k_i + 2))$ described by the equation $\sum_i X_i^{k_i+2}$ with its canonical polarization. In the case when $r = 5$ and $k_1 = \cdots = k_r = 3$, which corresponds to the Fermat quintic hypersurface in $\mathbb{P}^1$, he poins out the following facts in favor of this conjecture:

(i): The Fermat hypersurface Y_5 and the theory $E_{(3,\ldots,3)}$ have the same group of discrete symmetries.

(ii): The set of infinitesimal deformations of Y_5 endowed with a complexified Kähler parameter is isomorphic to that of $E_{(3,\ldots,3)}$, regarded as a representation of this discrete group.

4. Mirror symmetry

The Virasoro $N = 2$-superconformal algebra with central generator C has four series of odd generators G_s^+, G_s^-, $\overline{G}_s^+$, $\overline{G}_s^-$ with $s \in \mathbb{Z} + \frac{1}{2}$, corresponding to the Laurent expansions of the two series of supersymmetry generators, each being parameterized by a holomorphic or antiholomorphic spinor on $\mathbb{C}^*$. It admits the L_m and $\bar{L}_m$ as even generators for $m \in \mathbb{Z}$, corresponding to the Laurent expansions of holomorphic or antiholomorphic vector fields on $\mathbb{C}^*$. It also contains two series of even generators J_m for $m \in \mathbb{Z}$, called "the current $U(1)$" and representing the complex structure operator on X.

The charges $U(1)$ mentioned above are the eigenvalues of the operators J_0 and $\bar{J}_0$. A representation having central charge c sends C to $c\mathrm{Id}$. The (super)commutation relations are the following:

(a): all the brackets between barred and non-barred generators are trivial.

(b): $\{G_s^+, G_r^+\} = \{G_s^-, G_r^-\} = \{\overline{G}_s^+, \overline{G}_r^+\}, = \{\overline{G}_s^-, \overline{G}_r^-\} = 0$.

(c): The nontrivial relations are essentially the following:

(2.39)
$$\begin{cases} [L_m, L_n] = (m - n) L_{m+n} + \frac{1}{12} m(m^2 - 1) \delta_{m+n,0} C, \\ \{G_r^+, G_s^-\} = 2 L_{r+s} - (r - s) J_{r+s} + \frac{1}{3}\left(r^2 - \frac{1}{4}\right) \delta_{r+s,0} C, \end{cases}$$

and relations similar to (2.39) with the barred generators. The existence of the mirror phenomenon derives from the following observation: the superalgebra admits an involution:

(2.40)
$$\begin{cases} G_r^+ \mapsto G_r^-, & G_r^- \mapsto G_r^+, \\ \overline{G}_r^+ \mapsto \overline{G}_r^+, & \overline{G}_r^- \mapsto \overline{G}_r^-, \\ J_m \mapsto -J_m, & \bar{J}_m \mapsto \bar{J}_m. \end{cases}$$

This involution acts on the representations of the superalgebra, obviously producing isomorphic $N = 2$-superconformal theories. However, when the theory derives from the geometry, that is, when it is obtained by quantification of the $N = 2$-supersymmetric σ-model associated with a Calabi–Yau manifold endowed with a complexified Kähler parameter ω, the generators G^+, G^-, $\overline{G}^+$, and $\overline{G}^-$ have a precise geometric meaning (see (2.25) and (2.28)) and so cannot be interchanged. The mirror of (X, ω) should be precisely the couple (X', ω') corresponding (admitting the Gepner conjecture) to the conformal "mirror" theory obtained by appling the involution (2.40), isomorphic to the former, but with a different marking of the supersymmetry generators.

4.1. Examples. Greene and Plesser [31] have worked on Gepner's models (see 3), more precisely on the model $E_{(3,\dots,3)}$ corresponding to the Fermat quintic Y of degree 5 in $\mathbb{P}^4$. This hypersurface admits the symmetry group $(\mathbb{Z}/5\mathbb{Z})^5/\text{diag}$ acting on $\mathbb{P}^4$ by

(2.41) $$(\zeta_1, \dots, \zeta_5)^*(X_1, \dots, X_5) = (\zeta_1 X_1, \dots, \zeta_5 X_5),$$

where $\mathbb{Z}/5\mathbb{Z}$ is identified with the group of fifth roots of unity. The subgroup $G \subset (\mathbb{Z}/\mathbb{Z})^5/\text{diag}$ made up of the automorphisms of Y that act trivially on the type (3,0) form of Y is defined by the relation $\sum_i \alpha_i = 0$.

This group G acts on the associated conformal theory $E_{(3,\dots,3)}$, and for each subgroup $H \subset G$ one can form a superconformal theory $E_{(3,\dots,3)}^H$ constructed starting from the space of invariants in $E_{(3,\dots,3)}$ under H. The group G admits a nondegenerate bilinear form with values in $\mathbb{Z}/5\mathbb{Z}$ ($\langle \alpha, \beta \rangle = \sum_i \alpha_i \beta_i$) and thus, for each subgroup $H \subset G$ we have a subgroup $H^\perp \subset G$.

Greene and Plesser have shown by calculating the "partition functions" that $E_{(3,\dots,3)}^H$ and $E_{(3,\dots,3)}^{H^\perp}$ are mirrors of each other in the sense described above. This suggests that the quotients Y/H and $Y/H^\perp$, endowed with the canonical parameter given by $c_1(\mathcal{O}_Y(1))$ are mirrors of each other. This was established partially but rigorously by Roan [16], who has shown the equalities of the dimensions corresponding to (1.48).

THEOREM 2.6. *There exists a (canonical) desingularization $\widetilde{Y/H}$ of Y/H with trivial canonical bundle such that the following equalities hold:*

(2.42) $$h^1(T_{\widetilde{Y/H}}) = h^1(\Omega_{\widetilde{Y/H^\perp}}), \quad h^1(T_{\widetilde{Y/H^\perp}}) = h^1(\Omega_{\widetilde{Y/H}}).$$

5. The $N = 2$-superconformal theory and Dolbeault cohomology

Let $\mathcal{H}$ be the Hilbert space of the representation; we define (see [35]) finite-dimensional subspaces $R_{c,c}$ (resp. $R_{a,c}$) of $\mathcal{H}$, called (*chiral, chiral*) *state spaces* (resp. (*antichiral, chiral*) *state spaces*) as follows:

(2.43) $\quad R_{c,c} = \{x \in \mathcal{H}; G^+_{n-1/2}(x) = \overline{G}^+_{n-1/2}(x) = 0,$
$$G^-_{n+1/2}(x) = \overline{G}^-_{n-1/2}(x) = 0, \ n \geq 0\}$$

(2.44) $\quad R_{a,c} = \{x \in \mathcal{H}; G^+_{n+1/2}(x) = \overline{G}^+_{n-1/2}(x) = 0,$
$$G^-_{n-1/2}(x) = \overline{G}^-_{n+1/2}(x) = 0, \ n \geq 0\}.$$

We note that $R_{a,c}$ is obtained by interchanging G^+ and G^- in the definition of $R_{c,c}$, so that the involution (2.40) interchanges the spaces $R_{a,c}$ and $R_{c,c}$ (which means that the space $R_{c,c}$ of the initial theory is equal to the space $R_{a,c}$ of the theory obtained by applying (2.40)).

Using the rules of (super)commutation (2.39), we see that $R_{c,c}$ and $R_{a,c}$ are stable under J_0 and $\bar{J}_0$, which endows them with a bigrading, the eigenvalues of J_0 and $\bar{J}_0$ being assumed integers (see 2.3 and 2.4). The bidegrees for $R_{c,c}$ are of the form $(-p, q)$ with $p, q \in \mathbb{N}$ and the bidegrees for $R_{a,c}$ are of the form (p, q) with $p, q \in \mathbb{N}$.

If the central charge c is $3n$, it is shown in [**35**] that $R_{c,c}^{-n,n}$ is of rank 1, since $R_{a,c}^{n,n}$ is of rank 1. Since the theory is obtained by quantification of the $N = 2$-supersymmetric σ-model associated with a Calabi–Yau manifold X endowed with a complexified Kähler parameter ω, these spaces have the following interpretation

(2.45) $$R_{c,c} \cong \bigoplus_{p,q} H^q(\overset{p}{\wedge} T_X),$$

(2.46) $$R_{a,c} \cong \bigoplus_{p,q} H^q(\Omega^p_X),$$

where $H^q(\overset{p}{\wedge} T_X)$ has bidegree $(-p, q)$, $H^q(\Omega^p_X)$ has bidegree (p, q), and the isomorphisms preserve the bigradings.

Since the mirror involution (2.40) interchanges $R_{a,c}$ and $R_{c,c}$, these isomorphisms explain the series of isomorphisms (1.48) if the mirror map of § 1.8 corresponds to the involution (2.40) on the level of associated superconformal theories (see 2.4).

Intuitively these isomorphisms should be obtained as follows. As we are working on the $N = 2$-supersymmetric σ-model, the Hilbert space $\mathcal{H}$ should be a space of sections of the bundle on LX whose fiber at the point $\eta : S^1 \to X$ of LX is the space of $\mathcal{C}^\infty$ sections of $\eta^*(\Omega^\bullet_X)$.

By relations (2.39) the states annihilated by $G^+_{n+1/2}$, $G^-_{n+1/2}$, $\overline{G}^+_{n+1/2}$, and $\overline{G}^-_{n+1/2}$ for $n \geq 0$ are also annihilated by $\bar{L}_m$ and L_m for $m > 0$, which suggests that they are given by sections of $\Omega^\bullet_X$, since the L_m represent the infinitesimal action of Diff S^1 on $\mathcal{H}$. On the other hand, on these states the operators $G^+_{-1/2}$, $G^-_{-1/2}$, $\overline{G}^+_{-1/2}$, and $\overline{G}^-_{-1/2}$ can be identified with the operators ∂, ∂^*, $\bar{\partial}$, and $\bar{\partial}^*$ of X.

It is also shown in [**35**] that one can construct a bigraded commutative ring structure on each of the spaces $R_{c,c}$ and $R_{a,c}$. When combined with the (perhaps noncanonical) isomorphisms

$$R_{c,c}^{-n,n} \cong \mathbb{C}, \quad R_{a,c}^{n,n} \cong \mathbb{C}, \tag{2.47}$$

and the isomorphisms (2.45) and (2.46), they provide homogeneous forms of degree n on $H^1(\Omega_X)$ and on $H^1(T_x)$, called *Yukawa couplings* and denoted Y_1 and Y_2:

$$\begin{cases} Y_1 : S^n H^1(\Omega_X) \cong S^n R_{a,c} \longrightarrow R_{a,c}^{n,n} \cong \mathbb{C}, \\ Y_2 : S^n H^1(T_X) \cong S^n R_{c,c} \longrightarrow R_{c,c}^{-n,n} \cong \mathbb{C}. \end{cases} \tag{2.48}$$

The homogeneous forms constructed on $R_{a,c}$ and $R_{c,c}$ depend only on the assignments made in the conformal theory and are thus not changed by passage to the mirror theory. From that fact one can deduce that the mirror isomorphisms of (1.48)

$$H^1(T_X) \cong H^1(\Omega_{X'}), \quad H^1(T_X) \cong H^1(\Omega_X), \tag{2.49}$$

up to a coefficient, transform Y_2^X into $Y_1^{X'}$ and $Y_2^{X'}$ into Y_1^X.

6. Witten's interpretation

Witten [**38**] interprets mirror symmetry in a slightly more geometric way: the $N = 2$-supersymmetric σ-model of § 2.1.5 is described by the action $S(\phi, \psi)$ of (2.24) after elimination of the auxiliary parameters. By decomposing $\psi \in \mathcal{S} \otimes \phi^* T_X \otimes \mathbb{C}$ into holomorphic and antiholomorphic spinors, one can write

$$S(\phi, \psi) = S(\phi, \psi_+, \psi_-),$$

where $\psi_+ \in K^{1/2} \otimes \phi^* T_X$ and $\psi_- \in \overline{K}^{1/2} \otimes \phi^* T_X$.

When X is endowed with a complex structure, one can decompose $TX \otimes \mathbb{C}$ into $TX^{1,0} \oplus TX^{0,1}$. The action then becomes $S(\phi, \psi_+, \bar\psi_+, \psi_-, \bar\psi_-)$, where

$$\psi_+ \in K^{1/2} \otimes \phi^* TX^{1,0}, \quad \bar\psi_+ \in K^{1/2} \otimes \phi^* TX^{0,1},$$
$$\psi_- \in \overline{K}^{1/2} \otimes \phi^* T_X^{1,0}, \quad \bar\psi \in \overline{K}^{1/2} \otimes \phi^* T_X^{0,1}.$$

Witten then proposes to construct two models starting from this action by twisting the bundles in which ψ_+, $\bar\psi_+$, ψ_-, and $\bar\psi_-$ assume their values.

- In the A-model we take

$$\psi_+ \in \phi^* T_X^{1,0}, \quad \bar\psi_+ \in K \otimes \phi^* TX^{0,1},$$
$$\psi_- \in \overline{K} \otimes \phi^* T_X^{1,0}, \quad \bar\psi_- \in \phi^* T_X^{0,1}.$$

- In the B-model we take

$$\psi_+ \in K \otimes \phi^* T_X^{1,0}, \quad \bar\psi_+ \in \phi^* TX^{0,1},$$
$$\psi_- \in \overline{K} \otimes \phi^* T_X^{1,0}, \quad \bar\psi_- \in \phi^* T_X^{0,1}.$$

Then mirror symmetry would interchange the A- and B-models if we pass to the mirror manifold X'.

6.1. Yukawa couplings.

Witten calculates the Yukawa couplings Y_1 defined by Lerche–Vafa–Warner on $H^1(\Omega_X) \cong H^2(X, \mathbb{C})$ as correlation functions of the A-model:

$$(2.50) \quad Y_1(\omega_1, \ldots, \omega_n) = \int_{\phi, \psi, \chi} \mathcal{O}_1(p_1) \cdots \mathcal{O}_n(p_n) \exp\bigl(-S(\phi, \psi, \chi)\bigr) \, d\phi \, d\psi \, d\chi.$$

The symbols in this expression have the following meanings:

- $\chi = \psi_+ + \bar{\psi}_- \in \phi^* T_X$;
- ψ stands for the remaining fermionic variables;
- the p_i are points of $\mathbb{P}^1$;
- if, in local coordinates x_k on X, a closed 2-form representing the class ω_i is written

$$\sum_{k,l} \alpha_{kl} \, dx_k \wedge dx_l,$$

then the functional $\mathcal{O}_i(p_i)$ is defined by

$$(2.51) \quad \mathcal{O}_i(p_i)(\phi, \chi) = \sum_{kl} \alpha_{kl}\bigl(\phi(p_i)\bigr) \chi_k \chi_l;$$

- the integration is carried out on all the maps $\phi : \mathbb{P}^1 \to X$ (and the result is independent of the choice of the p_i).

The set of maps $\phi : \mathbb{P}^1 \to X$ splits into components corresponding to the homology classes $\phi_*\bigl([\mathbb{P}^1]\bigr) \in H_2(X, \mathbb{Z})$.

It will be recalled on the other hand that if ω is the Kähler form associated with the metric on X, the bosonic part of S can be written

$$(2.52) \quad S(\phi) = \int_\Sigma \phi^* \omega + \int_\Sigma 2\|\bar{\partial}\phi\|^2$$

where the first term depends only on the homology class $\phi_*\bigl([\mathbb{P}^1]\bigr)$, while the second term is positive and vanishes exactly on the holomorphic maps $\phi : \mathbb{P}^1 \to X$. The integral (2.50) can thus be written

$$(2.53) \quad \sum_{\alpha \in H_2(X, \mathbb{Z})} \exp\left(-\int_\alpha \omega\right) \int_{\phi_\alpha, \psi, \chi} \mathcal{O}_1(p_1) \cdots \mathcal{O}_n(p n) \, e^{-S'(\phi, \psi, \chi)} d\phi_\alpha \, d\psi \, d\chi$$

where ϕ_α ranges over the set of maps $\phi : \mathbb{P}^1 \to X$ such that $\phi_*\bigl([\mathbb{P}^1]\bigr) = \alpha$ and S' is the action obtained by suppressing the topological term $\int_\Sigma \phi^* \omega$ in S.

Witten then claims that the integral

$$\int_{\phi_\alpha, \psi, \chi} \mathcal{O}_1(p_1) \cdots \mathcal{O}_n(pn) \exp\bigl(-S'(\phi, \psi, \chi)\bigr) \, d\phi_\alpha \, d\psi \, d\chi$$

is, up to a coefficient, independent of the metric when the ω_i are closed, so that, if we replace g by tg and let t tend to $+\infty$, the bosonic part of S' being positive, this integral reduces to an integral over the set of ϕ_α that annihilates the bosonic part of S', that is, on the set of holomorphic maps $\phi : \mathbb{P}^1 \to X$ of class α. The following formula results:

$$(2.54) \quad Y_1(\omega_1, \ldots, \omega_n) = \int_X \omega_1 \cdots \omega_n + \sum_{\substack{\alpha \in H_2(X, \mathbb{Z}) \\ \alpha \neq 0}} \exp\left(-\int_\alpha \omega\right) n(\alpha) \int_\alpha \omega_1 \cdots \int_\alpha \omega_n$$

where $n(\alpha) \in \mathbb{Q}$ is an integral over the family of holomorphic maps of class α.

In Chapter 5 we shall discuss again the significance of the term $\int_\alpha \omega_1 \cdots \int_\alpha \omega_n$ in this formula and, in the case of dimension 3, the calculation of $n(\alpha)$.

The Yukawa couplings over $H^1(T_X)$ are computed as correlation functions of the model B, the observables corresponding then to (0,1)-forms with values in $T_X^{1,0}$. By similar arguments Witten shows that the functional integral then reduces to an integral over the set of constant maps $\phi : \mathbb{P}^1 \to X$, that is, on X. The formula is then the following

(2.55) $$Y_2(u_1, \ldots, u_n) = \langle \kappa^2, u_1 \cdots u_n \rangle_X,$$

where κ is a holomorphic section of K_X, hence defined up to a coefficient; $u_1 \cdots u_n$, which belongs to $H^n(\overset{n}{\wedge} T_X)$, is the cup-product of the classes $u_i \in H^1(T_X)$, and $\langle \, , \rangle_X$ is the perfect coupling $H^n(\overset{n}{\wedge} T_X) \otimes H^0(K_X^{\otimes 2}) \to H^n(K_X) = \mathbb{C}$.

CHAPTER 3

The Work of Candelas–de la Ossa–Green–Parkes

This chapter is devoted to variations of Hodge structure of weight 3. In the first part we show how the marked period map makes it possible to construct natural coordinates on the space of deformations of the marked complex structure of a Calabi–Yau threefold. We also describe the Yukawa couplings on the tangent spaces to these deformations in terms of infinitesimal variation of Hodge structure and show that, when normalized in a natural way, they depend on a potential in these coordinates.

We then explain, following Morrison, how these considerations extend to the boundary of the moduli space; and after sketching the proof of the theorem on quasiunipotence of the monodromy, we show that for a family of dimension 1 there exists a natural coordinate q in a neighborhood of a maximally unipotent degeneracy. This coordinate is given by the exponential of certain periods (determined by the monodromy action), which are solutions of a Picard–Fuchs equation. After defining the latter, we explain the computation of their coefficients for complete intersection families using the representation of the cohomology by residues of meromorphic differential forms.

We show finally how these techniques combine in [43] to give an integer series $\psi(q)$, calculating the normalized Yukawa couplings applied to the logarithmic field $q\partial/\partial q$ for the mirror family of the family of quintics of $\mathbb{P}^4$, and conclude, as in [43], that the identification of that series with the Gromov–Witten potential of the quintics makes it possible to predict for every d the (conjecturally finite) number of rational curves generically embedded of degree d in a general quintic.

1. Special coordinates and Yukawa couplings

In this chapter we are considering Calabi–Yau threefolds X with $b_1(X) = 0$. It is known by Theorem 1.9 that the universal local family B of deformations of X is smooth, of dimension $h^1(T_X) = h^1(\Omega_X^2)$.

1.1. Special coordinates on B. Let

$$\pi : \mathcal{X} \longrightarrow B, \quad X \cong X_0 := \pi^{-1}(0)$$

be the universal deformation of X parameterized by B. We assume that B is simply connected; we then have a canonical isomorphism:

$$R^3\pi_*(\mathbb{Z}) \cong H^3(X_0, \mathbb{Z}).$$

The group $H^3(X_0, \mathbb{Z})$ of integer cohomology modulo torsion is endowed with its natural intersection form $\langle\,,\,\rangle$, which is skew-symmetric and unimodular. Let $(\kappa)_{b \in B}$ be a nonzero holomorphic section at 0 of the line bundle $\mathcal{H}^{3,0}$ (Theorem 1.15), which means that $b \mapsto \kappa_b$ is a holomorphic function with values in $H^3(X_0, \mathbb{C})$, which can be canonically identified with $H^3(X_b, \mathbb{C})$, having the property that $\kappa_b \in H^{3,0}(X_b)$ for any point $b \in B$.

The following lemma shows that an adequate choice of a symplectic base of $H_3(X_0, \mathbb{Z})$ makes it possible, on the one hand to "normalize" $(\kappa_b)_{b \in B}$, and on the other hand to construct canonical coordinates on B.

LEMMA 3.1. *For a suitable choice of* $(\alpha_0, \ldots, \alpha_N, \beta_0, \ldots, \beta_N)$, $N = h^1(\Omega_X^2)$, *forming a symplectic base of* $H_3(X_0, \mathbb{Z})$, *that is, for which the form* $\langle , \rangle$ *satisfies* $\langle \alpha_i, \alpha_j \rangle = \langle \beta_i, \beta_j \rangle = 0$ *and* $\langle \alpha_i, \beta_j \rangle = \delta_i^j$, *we have, in a neighborhood of* 0,
$$\int_{\alpha_0} \kappa_b \neq 0,$$
so that one can determine a section κ' *of* $\mathcal{H}^{3,0}$ *by the condition* $\int_{\alpha_0} \kappa'_b = 1$, *and the functions* $z_i(b) = \int_{\alpha_i} \kappa'_b$ *for* $i > 0$ *provide coordinates on* B *in a neighborhood of* 0.

PROOF. One can find a primitive element $\alpha_0 \in H_3(X_0, \mathbb{Z})$ such that $\int_{\alpha_0} \kappa_0 \neq 0$ since $\kappa_0 \neq 0$. We therefore have $\int_{\alpha_0} \kappa_b \neq 0$ in a neighborhood of 0, and the section κ' is defined by $\kappa'_b = \kappa_b / \int_{\alpha_0} \kappa_b$. For a symplectic base $(\alpha_0, \ldots, \beta_N)$ as above, extending α_0, we consider the map $\phi : B \to \mathbb{C}^N$ given by $\phi^* z_i = \int_{\alpha_i} \kappa'$. Its differential at 0 is given by the composition

$$(3.1) \qquad T_{B,0} \xrightarrow{d(\kappa')_0} H^3(X_0, \mathbb{C}) \xrightarrow{\int_{\alpha_i}} \mathbb{C}^N.$$

By Theorem 1.17, the map $d(\kappa')_0$, modulo $\mathcal{H}^{3,0}(X_0)$, is given by the composition

$$(3.2) \qquad T_{B,0} \cong H^1(T_{X_0}) \stackrel{\kappa'_0}{\cong} H^1(\Omega_{X_0}^2) \subset H^3(X_0, \mathbb{C})/H^{3,0}(X_0).$$

Since $\int_{\alpha_0} \big(d(\kappa')_0(T_{B_0}) \big) = 0$, $d\phi_0$ is not an isomorphism if and only if there exists $\eta \in F^2 H^3(X_0)$ such that $\int_{\alpha_i} \eta = 0$ for $i = 0, \ldots, N$, which is equivalent to the fact that η belongs to the subspace generated over $\mathbb{C}$ by the α_i (where we have used the Poincaré duality $H_3 \cong H^3$).

On $F^2 H^3(X_0)$ we consider the Hermitian form
$$h(u, v) = i \langle u, \bar{v} \rangle.$$
According to 1.5.3 (*ii*), it is nondegenerate and has signature $(+, -, \cdots, -)$. If we have η as above, then η satisfies $i \langle \eta, \bar{\eta} \rangle = 0$ (since $\langle \alpha_i \rangle$ is totally isotropic and stable under conjugation) and $\int_{\alpha_0} \eta = 0$. This is possible only if the projection α'_0 of α_0 on $F^2 H^3$ (parallel to $\overline{F^2 H^3}$) satisfies $h(\alpha'_0, \alpha'_0) \leq 0$. But the projection of $H_3(X_0, \mathbb{Z}) \cong H^3(X_0, \mathbb{Z})$ on $F^2 H^3$ is a lattice, and thus cannot be contained in the set defined by the condition $h(\eta, \bar{\eta}) \leq 0$. If we have chosen an element α_0 of $H_3(X_0, \mathbb{Z})$ satisfying $h(\alpha'_0, \alpha'_0) > 0$ and $\int_{\alpha_0} \kappa_0 \neq 0$, the preceding shows that $d\phi_0$ is an isomorphism and that the z_i provide coordinates on B in a neighborhood of 0. $\square$

1.2. Yukawa couplings. It will be recalled that the Yukawa couplings Y_2 (see (2.55)) on $H^1(T_{X_0}) \cong T_{B_0}$ are given by the cubic form

$$(3.3) \qquad Y_2 : S^3 H^1(T_{X_0}) \to \mathbb{C}$$

depending on the choice of $\kappa \in H^{3,0}(X_0)$ defined by

(3.4) $$Y_2(u_1, u_2, u_3) = \langle \kappa^2, u_1 \cdot u_2 \cdot u_3 \rangle,$$

where $u_1 \cdot u_2 \cdot u_3 \in H^3(\overset{3}{\wedge} T_{X_0})$ and $\langle , \rangle$ is given by Serre duality.

We shall use the notation Y_2^κ for the coupling Y_2 defined by the form κ.

By considering, as above, $(\kappa_b)_{b \in B}$ as a holomorphic map $\kappa : B \to H^3(X_0, \mathbb{C})$, we shall show the following.

LEMMA 3.2. *Let $(\kappa_b)_{b \in B}$ be a holomorphic section of $\mathcal{H}^{3,0}$, and let x_i be coordinates on B. Then*

(3.5) $$Y_2^{\kappa_0}\left(\frac{\partial}{\partial x_i}, \frac{\partial}{\partial x_j}, \frac{\partial}{\partial x_k}\right) = \left\langle \kappa_0, \frac{\partial^3 \kappa}{\partial x_i \partial x_j \partial x_k}(0) \right\rangle_{H^3(X_0, \mathbb{C})}.$$

The lemma follows from the transversality, the polarization conditions 1.5.3 (i), and Theorem 1.17. We have indeed

$$\left\langle \kappa_0, \frac{\partial^3 \kappa}{\partial x_i \partial x_j \partial x_k}(0) \right\rangle_{H^3(X_0, \mathbb{C})} = \left\langle \kappa_0, \left(\frac{\partial^3 \kappa}{\partial x_i \partial x_j \partial x_k}(0)\right)^{0,3} \right\rangle,$$

where $(\eta)^{0,3}$ is the projection of η on $H^{0,3}(X_0) \cong H^3(\mathcal{O}_{X_0})$ and the bracket $\langle , \rangle$ denotes the Serre duality between $H^0(K_{X_0})$ and $H^3(\mathcal{O}_{X_0})$. On the other hand, by repeated application of Theorems 1.16 and 1.17 we see that

(3.6) $$\left(\frac{\partial^3 \kappa}{\partial x_i \partial x_j \partial x_k}(0)\right)^{0,3} = \kappa_0 \cdot \frac{\partial}{\partial x_i} \cdot \frac{\partial}{\partial x_j} \cdot \frac{\partial}{\partial x_k} \in H^3(\mathcal{O}_{X_0}),$$

where $\partial/\partial x_i \cdot \partial/\partial x_j \cdot \partial/\partial x_k \in H^3(\wedge^3 T_{X_0}) \cong H^3(K_{X_0}^{-1})$. Finally, it suffices to note the following equality for $\kappa \in H^0(K_{X_0})$ and $\chi \in H^3(K_{X_0}^{-1})$

(3.7) $$\langle \kappa, \kappa \cdot \chi \rangle = \langle \kappa^2, \chi \rangle,$$

where the first bracket is the Serre duality between $H^0(K_{X_0})$ and $H^3(\mathcal{O}_{X_0})$ and the second bracket is the Serre duality between $H^0(K_{X_0}^{\otimes 2})$ and $H^3(K_{X_0}^{-1})$. □

A remarkable property of the special coordinates and the normalization of $(\kappa_b)_{b \in B}$ constructed above is the following fact.

PROPOSITION 3.3. *Let $\{\alpha_0, \ldots, \beta_N\}$ be a symplectic base of $H_3(X_0, \mathbb{C})$, making it possible to construct the normalized section κ' of $\mathcal{H}^{3,0}$ and the coordinates z_i as in Lemma 3.1. Then the Yukawa couplings $Y_2^{\kappa'}(\partial/\partial z_i, \partial/\partial z_j, \partial/\partial z_k)$ depend on a potential, that is, there exists a function $F(z_1, \ldots, z_n)$ such that*

(3.8) $$Y_2^{\kappa'}\left(\frac{\partial}{\partial z_i}, \frac{\partial}{\partial z_j}, \frac{\partial}{\partial z_k}\right) = \frac{\partial^3 F}{\partial z_i \partial z_j \partial z_k}.$$

The proposition follows from Lemmas 3.4 and 3.5 below. Let

$$\phi_i(z_1, \ldots, z_n) = \int_{\beta_i} \kappa', \quad i = 1, \ldots, n,$$

where κ' is regarded as a function of (z_i), since the z_i are coordinates on B.

LEMMA 3.4. *The relation $\dfrac{\partial \phi_i}{\partial z_j} = \dfrac{\partial \phi_j}{\partial z_i}$ holds.*

This lemma follows from the fact that $\operatorname{Im}(d\kappa')_b \subset F^2 H^3(X_b)$ is totally isotropic for the intersection form $\langle\, ,\,\rangle$ on $H^3(X_0, \mathbb{C})$. Indeed, by definition of z_i and ϕ_i and the fact that $\int_{\alpha_0} \kappa' = 1$, we have:

$$(3.9) \qquad \kappa' = \beta_0 + \sum_{j>0} z_j \beta_j - \sum_{j>0} \phi_j \alpha_j - \left(\int_{\beta_0} \kappa'\right) \alpha_0.$$

We thus obtain:

$$(3.10) \qquad \frac{\partial \kappa'}{\partial z_i} = \beta_i - \sum_{j>0} \frac{\partial \phi_j}{\partial z_i} \alpha_j - \frac{\partial \left(\int_{\beta_0} \kappa'\right)}{\partial z_i} \alpha_0.$$

The equality $\langle \partial \kappa'/\partial z_i, \partial \kappa'/\partial z_j \rangle = 0$, which is due to the polarization conditions and the fact that each of the sections is in $F^2 \mathcal{H}^3$ by transversality, then gives the lemma immediately. $\square$

Thus there exists a function $F(z_1, \ldots, z_n)$ such that $\phi_i = \partial F/\partial z_i$.

LEMMA 3.5. *The relation* $\dfrac{\partial^3 F}{\partial z_i \partial z_j \partial z_k} = -Y_2^{\kappa'}\left(\dfrac{\partial}{\partial z_i}, \dfrac{\partial}{\partial z_j}, \dfrac{\partial}{\partial z_k}\right)$ *holds.*

Indeed, we note that by transversality $\dfrac{\partial^2 \kappa'}{\partial z_i \partial z_j}$ belongs to $F^1 H^3(X_b) = H^{3,0}(X_b)^\perp$ and hence

$$(3.11) \qquad \left\langle \kappa', \frac{\partial^2 \kappa'}{\partial z_i \partial z_j} \right\rangle = 0.$$

This gives

$$(3.12) \qquad \left\langle \frac{\partial \kappa'}{\partial z_k}, \frac{\partial^2 \kappa'}{\partial z_i \partial z_j} \right\rangle + \left\langle \kappa', \frac{\partial^3 \kappa'}{\partial z_i \partial z_j \partial z_k} \right\rangle = 0.$$

By virtue of Lemma 3.2 we thus have:

$$Y_2^{\kappa'}\left(\frac{\partial}{\partial z_i}, \frac{\partial}{\partial z_j}, \frac{\partial}{\partial z_k}\right) = -\left\langle \frac{\partial \kappa'}{\partial z_k}, \frac{\partial^2 \kappa'}{\partial z_i \partial z_j} \right\rangle.$$

On the other hand, by (3.10) we have

$$(3.13) \qquad \left\langle \frac{\partial \kappa'}{\partial z_k}, \frac{\partial^2 \kappa'}{\partial z_i \partial z_j} \right\rangle = \left\langle \beta_k - \sum_{l>0} \frac{\partial \phi_l}{\partial z_k} \alpha_l - \frac{\partial}{\partial z_k}\left(\int_{\beta_0} \kappa'\right) \alpha_0, \right.$$
$$\left. -\left(\sum_{l>0} \frac{\partial^2 \phi_l}{\partial z_i \partial z_j} \alpha_l + \frac{\partial^2}{\partial z_i \partial z_j}\left(\int_{\beta_0} \kappa'\right) \alpha_0\right) \right\rangle$$
$$= \frac{\partial^2 \phi_k}{\partial z_i \partial z_j} = \frac{\partial^3 F}{\partial z_i \partial z_j \partial z_k},$$

which proves Lemma 3.5 and hence Proposition 3.3. $\square$

Proposition 3.3 is a partial mathematical justification of mirror symmetry. Indeed, if mirror symmetry exists, one has a local identification of the space of deformations of the complex structure of X with that of the deformations of the complexified Kähler parameter of its mirror, which, being the quotient by a group of translations of an open set in a vector space, possesses a canonical flat structure (that is, coordinates defined up to a linear transformation).

The Yukawa couplings Y_1 (see (2.54)), on the other hand, depend clearly on a potential, so that Proposition 3.3 provides in a natural way part of the structure predicted by mirror symmetry, assuming that the mirror map does indeed identify the natural flat structures of these two spaces.

We note finally that the system of coordinates and more generally the flat structure provided by Lemma 3.1 depend on the choice of the symplectic base (actually it latter depends only on the assignment of α_0 and on the totally isotropic subspace generated by the α_i), which indicates that mirror symmetry is well-defined only on spaces of marked complex structure.

2. Degenerations

2.1. Extension of the Hodge filtration. Let $\pi : \mathcal{X} \to B$ be a proper flat map with $\mathcal{X}$ smooth and $\dim B = 1$, and let $0 \in B$ be such that $X := \pi^{-1}(0)$ is singular. Shrinking B to a neighborhood of 0, one can then assume that π is smooth over $B^* = B - \{0\}$. On B^* we have the bundle $\mathcal{H}^n$ endowed with its Hodge filtration $F^{\bullet}\mathcal{H}^n$ (see 5). the following theorem is due to Katz [**50**] and, in a more general context, to Schmidt [**56**].

THEOREM 3.6. *The bundle $\mathcal{H}^n$ over B^* can be extended to a bundle $\overline{\mathcal{H}}^n$ over B, such that the Gauss–Manin connection ∇ extends to a connection having logarithmic singularities*

$$(3.14) \qquad \nabla : \overline{\mathcal{H}}^n \to \overline{\mathcal{H}}^n \otimes \Omega_B(\log 0)$$

and such that the Hodge filtration extends to a filtration $F^{\bullet}\overline{\mathcal{H}}^n$ of $\overline{\mathcal{H}}^n$ by subbundles.

By continuity, the property of transversality (Theorem 1.16) remains true for the extended Hodge filtration:

$$(3.15) \qquad \nabla(F^p\overline{\mathcal{H}}^n) \subset F^{p-1}\overline{\mathcal{H}}^n \otimes \Omega_B(\log 0).$$

In the geometric case that we are considering, this extension is obtained as follows (see [**57**], [**58**]).

Taking a covering of B branched at 0, and performing birational transformations along the central fiber, we can assume that X is a reduced divisor having normal crossings (Mumford's semi-stable reduction theorem). We then consider on $\mathcal{X}$ the complex $\Omega^{\bullet}_{\mathcal{X}/B}(\log X)$ defined as follows. For a coordinate t on B centered at 0 and functions $z_1, \ldots, z_r$ with independent differentials on $\mathcal{X}$, π is described locally by

$$(3.16) \qquad \pi^*(t) = z_1 \cdots z_r.$$

We call

$$\Omega_{\mathcal{X}/B}(\log X)$$

the quotient of the sheaf *generated* by $\Omega_{\mathcal{X}}$ and the dz_i/z_i by the relation

$$\pi^*\left(\frac{dt}{t}\right) = \sum_i \frac{dz_i}{z_i} = 0,$$

and we set:

$$\Omega^k_{\mathcal{X}/B}(\log X) = \overset{k}{\wedge} \Omega_{\mathcal{X}/B}(\log X).$$

We have the exterior differential $d : \Omega^\bullet_{\mathcal{X}/B}(\log X) \to \Omega^{\bullet+1}_{\mathcal{X}/B}(\log X)$; hence $\Omega^\bullet_{\mathcal{X}/B}(\log X)$ is a complex on $\mathcal{X}$. We then define the extension of the Hodge bundle by

$$(3.17) \qquad \overline{\mathcal{H}}^n = \mathbb{R}^n \pi_* \big(\Omega^\bullet_{\mathcal{X}/B}(\log X) \big)$$

and the extension of the Hodge filtration by

$$(3.18) \qquad F^p \overline{\mathcal{H}}^n = \mathbb{R}^n \pi_* \big(0 \to \Omega^p_{\mathcal{X}/B}(\log X) \to \cdots \to \Omega^n_{\mathcal{X}/B}(\log X) \to 0 \big).$$

Now suppose that the fibers X_b are manifolds with trivial canonical bundle of dimension n. Theorem 3.6 makes it possible to extend to B the Yukawa couplings Y_2 defined on B^*. Indeed, the condition of transversality (3.15) makes it possible to define $\mathcal{O}_B$-linear maps

$$(3.19) \qquad \phi_p : TB(\log 0) \to \mathrm{Hom}\,(F^p/F^{p+1}\overline{\mathcal{H}}^n, F^{p-1}/F^p\overline{\mathcal{H}}^n)$$

where $T_B(\log 0)$, which is by definition the dual of the sheaf $\Omega_B(\log 0)$ of holomorphic forms on B^* with logarithmic singularities at 0, is generated in a neighborhood of 0 by the field $t\partial/\partial t$. By composing the maps ϕ_p we then construct an arrow

$$(3.20) \qquad S^n TB(\log 0) \xrightarrow{\psi} \mathrm{Hom}\,(\overline{\mathcal{H}}^{n,0}, \overline{\mathcal{H}}^{0,n}).$$

But the definition of the filtration $F^p \overline{\mathcal{H}}^n$ and the degeneracy at E_1 of the Hodge–de Rham spectral sequence for the complex of sheaves $\Omega^\bullet_{\mathcal{X}/B}(\log X)$ show that $\overline{\mathcal{H}}^{n,0} = R^0\pi_* K_{\mathcal{X}/B}$ and $\overline{\mathcal{H}}^{0,n} = R^n\pi_* \mathcal{O}_{\mathcal{X}}$ are dual, of rank 1 in our case. We choose a nonzero section $(\kappa_b)_{b\in B}$ of $\overline{\mathcal{H}}^{n,0}$. We have thus a normalized Yukawa coupling:

$$(3.21) \qquad \begin{cases} Y_2^\kappa : S^n TB(\log 0) \longrightarrow \mathcal{O}_B, \\ (t\partial/\partial t)^{\otimes n} \mapsto \langle \kappa, \psi((t\partial/\partial t)^{\otimes n})(\kappa) \rangle. \end{cases}$$

We now consider, as in [43] and [53], families of Calabi–Yau threefolds with $h^{2,1} = 1$; we assume that we have a family of dimension 1 that is a Zariski open set Δ^* of a curve Δ parameterizing a family $\mathcal{X} \to \Delta$ with singular fiber over $\Delta - \Delta^*$. Lemma 3.1 shows the local existence of a privileged normalization $\kappa \in \mathcal{H}^{3,0}$ and a special coordinate z on the open set Δ^*, depending on the choice of a symplectic base of $H^3(X_t, \mathbb{Z})$, when Δ^* is, in a neighborhood of each of its points t, the local universal family of the fiber X_t.

As the base Δ^* is of dimension 1, the choice of a coordinate z and a section k of $\mathcal{H}^{3,0}$ identifies the Yukawa couplings with a function $\Phi^* = Y_2^\kappa(\partial/\partial z, \partial/\partial z, \partial/\partial z)$ on Δ^*. The considerations developed in 2.1 show that one can also construct such a function Φ in a neighborhood of $0 \in \Delta - \Delta^*$. Indeed the choice of a coordinate q on Δ in a neighborhood of 0 and a section $\kappa \in \overline{\mathcal{H}}^{3,0}$ makes it possible to define

$$(3.22) \qquad \Phi = Y_2^\kappa\Big(q\frac{\partial}{\partial q}, q\frac{\partial}{\partial q}, q\frac{\partial}{\partial q}\Big).$$

We can now show, following [53] that under a convenient hypothesis on the monodromy of the local system $H^3_{\mathbb{Z}}$ around 0, there is a canonical choice of κ_b^2 and the coordinate q. An essential role is played by the (mixed) Hodge theory and the monodromy theorem recalled in the following paragraph.

2.2. Mixed Hodge theory and monodromy.

Let $\pi : \mathcal{X} \to \Delta$ be a family of projective varieties (by that we mean that there exists an embedding of $\mathcal{X}$ in $\Delta \times \mathbb{P}^K$ over Δ), where Δ is a disk, and π is smooth over Δ^*. We denote the central fiber by X.

Let b be a point of Δ^*; the group $\pi_1(\Delta^*, b) \cong \mathbb{Z}$ acts on $H^n(X_b, \mathbb{Z})$ through the diffeomorphism of X_b obtained by $\mathcal{C}^\infty$ trivialization of the family $\mathcal{X}$ over the universal covering of Δ^*. The monodromy endormorphism $T \in \mathrm{Aut}\,(H^n(X_b, \mathbb{Z}))$ is defined as the image of the canonical generator of $\pi_1(\Delta^*, b)$.

THEOREM 3.7. *The endomorphism T is quasiunipotent, which means that there exists k and $m \in \mathbb{N}$ such that $(T^k - 1)^m = 0$.*

A proof of this theorem due to Borel [48] uses the following fact, which is a consequence of the calculation of the curvature of the domain of polarized periods "in horizontal directions."

Let $\mathbb{H}$ be the upper half-plane endowed with its hyperbolic metric. Let $\mathcal{D}$ be the domain of polarized periods constructed on a real vector space H endowed with a bilinear form $\langle\,,\,\rangle$ (see 5.3), an integer k (the level) and the Hodge numbers $h^{p,q}$ with $p + q = k$ satisfying $\sum_p h^{p,q} = \dim H$ being given. The space $\mathcal{D}$ is a homogeneous space under the group $G = \mathrm{Aut}\,(H, \langle\,,\,\rangle)$ and thus admits a G-invariant metric. We then have for a correct normalization of the metric on $\mathcal{D}$:

PROPOSITION 3.8. *Let $\mathcal{P} : \mathbb{H} \to \mathcal{D}$ be a horizontal holomorphic map (that is, satisfying the condition of transversality). Then $\mathcal{P}$ decreases distances, that is*

$$(3.23) \qquad d_\mathcal{D}\big(\mathcal{P}(x), \mathcal{P}(y)\big) \leq d_\mathbb{H}(x, y).$$

The proposition implies the theorem in the following way: $\mathbb{H}$ is the universal covering of Δ^* and on $\mathbb{H}$ the local system $(H^n_\mathbb{Z})_{\mathrm{prim}}$ is trivial, which provides a polarized vector space $H = H^n_{\mathrm{prim}}(X_t, \mathbb{R})$ for $t \in \mathbb{H}$. (It is important to note that by the Lefschetz decomposition (1.38), it suffices to prove the theorem for the monodromy action on the primitive cohomology $(H^n_\mathbb{Z})_{\mathrm{prim}}$ relative to the polarization provided by the projective submersion of $\mathcal{X}$.) We thus have a map of polarized periods $\mathcal{P} : \mathbb{H} \to \mathcal{D}$ obtained by considering the variation of Hodge structure on the primitive cohomology of degree n of the induced family $\mathcal{X}_\mathbb{H}$. It is known that $\mathcal{P}$ decreases distances and satisfies

$$(3.24) \qquad \mathcal{P}(z+1) = T\mathcal{P}(z),$$

where T acts on $\mathcal{D}$ as in 5.4. But in $\mathbb{H}$ we have $\lim_{n\to\infty} d_\mathbb{H}(z_n, z_n + 1) = 0$ when $\lim_{n\to\infty} \mathrm{Im}\, z_n = \infty$. For a sequence z_n satisfying this condition, we thus find:

$$(3.25) \qquad \lim_{n\to\infty} d_\mathcal{D}\big(\mathcal{P}(z_n), T\mathcal{P}(z_n)\big) = 0.$$

Writing $\mathcal{P}(z_n) = g_n \mathcal{P}(z_0)$ with $g_n \in G$, and using the invariance of the metric $d_\mathcal{D}$ under G, we find

$$(3.26) \qquad \lim_{n\to\infty} d_\mathcal{D}\big(\mathcal{P}(z_0), g_n^{-1} T g_n \mathcal{P}(z_0)\big) = 0.$$

This shows that the adherence of the sequence $g_n^{-1} T g_n$ lies in the closure of the stabilizer of $\mathcal{P}(z_0)$ in G, which is compact by the polarization conditions 1.5.3 (*ii*). The eigenvalues of T are thus of absolute value 1, and as T is defined over $\mathbb{Z}$, the eigenvalues of T are roots of unity. □

In fact, there exists a more precise statement using the fact that, if we assume T unipotent, which we can do after a change of base of degree k, $N = \log T$ is a morphism of mixed Hodge structure, for the mixed Hodge structure on the limit fiber such that $N(W_k H^n) \subset W_{k-2} H^n$.

Since we shall need this mixed Hodge structure below, we explain briefly following [57] how one can construct it on $H_{\lim}^n = \overline{\mathcal{H}}_0^n$.

A *mixed Hodge structure* on a $\mathbb{Q}$-vector space is the assignment of an increasing filtration $W_\bullet V$ defined on $\mathbb{Q}$ and a decreasing (Hodge) filtration $F^\bullet V$ inducing a pure Hodge structure of weight k on the graded ring Gr_k^W associated with W.

The Hodge filtration on $\overline{\mathcal{H}}_0^n$ is described in (3.18).

Let $\Omega_{\mathcal{X}}(\log X)$ be the bundle on $\mathcal{X}$ generated locally by $\Omega_{\mathcal{X}}$ and the dz_i/z_i with the notation of (3.16) (where we have used the relative version $\Omega_{\mathcal{X}/B}(\log X)$).

Let $\Omega_{\mathcal{X}}^k(\log X) = \wedge^k \Omega_{\mathcal{X}}(\log X)$ and

$$(3.27) \qquad W_k' \Omega_{\mathcal{X}}^p(\log X) = \Omega_{\mathcal{X}}^k(\log X) \wedge \Omega_{\mathcal{X}}^{p-k} \subset \Omega_{\mathcal{X}}^p(\log X).$$

We construct the complex $B^k = \sum_{p+q=k} A^{p,q}$ associated with the double complex

$$(3.28) \qquad A^{p,q} = \Omega_{\mathcal{X}}^{p+q+1}(\log X) / W_q' \Omega_{\mathcal{X}}^{p+q+1}(\log X)$$

endowed with the differentials

- $d : A^{p,q} \to A^{p+1,q}$ (given by the exterior derivative) and
- $d' : A^{p,q} \to A^{p,q+1}$ (induced by the exterior product with $\pi^* dt/t$).

One can show that the inclusion

$$(3.29) \qquad \Omega_{\mathcal{X}/B}^p(\log X)_{|X} \subset A^{p,0}$$

given by the exterior product with $\pi^*(dt/t)$ gives a quasi-isomorphism

$$(3.30) \qquad \Omega_{\mathcal{X}/B}^\bullet(\log X)_{|X} \xrightarrow[\sim]{\mathrm{qis}} B^\bullet.$$

The filtration $W_k H_{\lim}^n$ is then induced by the following filtration on $A^{\bullet\bullet}$:

$$(3.31) \qquad W_k A^{p,q} = W_{2q+k-n+1}' \Omega_{\mathcal{X}}^{p+q+1}(\log X) / W_q' \Omega_{\mathcal{X}}^{p+q+1}(\log X).$$

It is clear that $W_{2n} H_{\lim}^n = H_{\lim}^n$ and $W_{-1} H_{\lim}^n = \{0\}$. On the other hand, one can show that $N(W_k H_{\lim}^n) \subset W_{k-2} H_{\lim}^n$ by noting that N can be identified up to a constant coefficient with the residue of the logarithmic connection (3.14) and that the latter is given by the short exact sequence

$$(3.32) \qquad 0 \to \Omega_{\mathcal{X}/B}^\bullet(\log X) \otimes \Omega_B(\log 0) \longrightarrow \Omega_{\mathcal{X}}^{\bullet+1}(\log X) \longrightarrow \Omega_{\mathcal{X}/B}^{\bullet+1}(\log X) \to 0,$$

which provides the relation

$$R^n \pi_* \left(\Omega_{\mathcal{X}/B}^\bullet(\log X) \right) \longrightarrow R^n \pi_* \left(\Omega_{\mathcal{X}/B}^\bullet(\log X) \right) \otimes \Omega_B(\log 0).$$

These two facts imply immediately:

THEOREM 3.9. *The monodromy endomorphism T is quasiunipotent of order $n+1$, that is, it satisfies $(T^k - 1)^{n+1} = 0$ on $H^n(X_b, \mathbb{Z})$.*

Notice that, in the case of Calabi–Yau threefolds with $h^{2,1} = 1$ this more refined statement follows immediately from Theorem 3.7, since the rank of $H^3(X, \mathbb{Z})$ equals 4.

2.3. Normalization of κ and choice of coordinates.

We now return to the case of a family $\pi : \mathcal{X} \to \Delta$ of Calabi–Yau threefolds with $h^{2,1} = 1$ and we suppose that Δ is a disk centered at 0 and that X_b is smooth for $b \neq 0$. We make the following assumption: the monodromy endomorphism $T \in \mathrm{Aut}\,(H^3(X_b, \mathbb{Z}))$ is *maximally unipotent*, that is:

$$(T-1)^4 = 0, \quad (T-1)^3 \neq 0.$$

Equivalently,

$$N = \log T = -\big((1-T) + \cdots + \tfrac{1}{3}(1-T)^3\big)$$

satisfies $N^4 = 0$ and $N^3 \neq 0$. We thus have, for an element $x \in H^3(X_b, \mathbb{Q})$,

$$H^3(X_b, \mathbb{Q}) = \langle x, Nx, N^2 x, N^3 x \rangle$$

and the image of N^3 is of rank 1. The group $\mathrm{Im}\,N^3 \cap H^3(X_b, \mathbb{Z})$ is thus generated by an element e_0 determined up to sign. As e_0 is annihilated by N, it is invariant under T. If κ is a holomorphic section of the bundle $\overline{\mathcal{H}}^{3,0}$, one can thus define on $B^* \cong \Delta^*$ a function

$$(3.33) \qquad g_0(z) = \langle e_0, \kappa_z \rangle_{H^3(X_z, \mathbb{C})}.$$

In fact g_0 extends holomorphically on Δ, since it is single-valued and its rate of growth is bounded by a power of $|\log z|$. We then have:

LEMMA 3.10. *If $\kappa_0 \neq 0$ in $\overline{\mathcal{H}}^{3,0}_0$, we have $g_0(0) \neq 0$.*

This results from the existence of a mixed Hodge structure on the limit fiber H^3_{lim} whose construction was sketched above.

For the filtration by the weight $W_\bullet H^3_{\mathrm{lim}}$, the graded group $\mathrm{Gr}^W_i H^3_{\mathrm{lim}}$ is endowed with a pure Hodge structure of weight i for $0 \leq i \leq 6$, induced by the Hodge filtration on H^3_{lim} (see (3.18)), and we have:

$$(3.34) \qquad \begin{cases} N^k(W_i H^3_{\mathrm{lim}}) \subset W_{i-2k} H^3_{\mathrm{lim}}, \\ N^k : \mathrm{Gr}^W_{3+k} H^3_{\mathrm{lim}} \cong \mathrm{Gr}^W_{3-k} H^3_{\mathrm{lim}} \end{cases}$$

where the isomorphism is an isomorphism of Hodge structures of degree $-2k$. From that one can deduce easily that

$$W_0 = \mathrm{Im}\,N^3 = \mathrm{Ker}\,N, \quad W_5 = \mathrm{Ker}\,N^3 = \mathrm{Im}\,N.$$

Since the Hodge structure on $\mathrm{Gr}^W_6 H^3_{\mathrm{lim}}$ is pure of weight 6 and rank 1, it is purely of type (3, 3), which implies that $\overline{\mathcal{H}}^{3,0}_0 \not\subset W_5 H^3_{\mathrm{lim}} = \mathrm{Ker}\,N^3$. Since $N = \log T$ satisfies the condition $\langle Nx, y \rangle = -\langle x, Ny \rangle$, we have $\mathrm{Ker}\,N^3 = (\mathrm{Im}\,N^3)^\perp$ (on $\mathbb{Q}$), and hence $\langle \kappa_0, e_0 \rangle \neq 0$, which proves the lemma. $\square$

Since $\mathrm{Im}\,N^2$ is of rank 2, there exists a vector $e_1 \in \mathrm{Im}\,N^2 \cap H^3(X_b, \mathbb{Z})$, which generates $(\mathrm{Im}\,N^2 \cap H^3(X_b, \mathbb{Z}))/e_0$. The element e_1 is determined up to a transformation $e_1 \mapsto \pm e_1 + k e_0$. Consider the function

$$(3.35) \qquad g_1(z) = \langle e_1, \kappa_z \rangle.$$

This is a multivalued function on B of growth bounded by a power of $|\log z|$ and its monodromy around 0 equals

$$g_1(e^{2i\pi} z) - g_1(z) = \langle Te_1, \kappa_z \rangle - \langle e_1, \kappa_z \rangle.$$

But, since e_1 is in $\operatorname{Im} N^2$, we have $Te_1 = (1+N)e_1$. There exists an integer $m \in \mathbb{Z}$ such that $Ne_1 = me_0$, and we obtain

(3.36) $$g_1(e^{2i\pi}z) - g_1(z) = mg_0(z).$$

The function
$$\frac{g_1(z)}{mg_0(z)} - \frac{1}{2i\pi}\log z$$
is then single-valued and has growth bounded by a power of $|\log z|$, hence is holomorphic on Δ. The function $t = g_1(z)/mg_0(z)$ therefore provides a coordinate for the universal covering of Δ^*, defined up to an additive rational constant k/m. If the condition

(∗) $$m = \pm 1$$

is satisfied, t is defined up to an additive integer constant, and $q = e^{2i\pi t}$ gives a canonical coordinate on B in a neighborhood of 0.

Modulo the condition (∗), we have thus shown that there exists a canonical normalization of κ^2 given by

(3.37) $$\kappa'^2 = \frac{\kappa^2}{g_0(x)^2}$$

and a canonical coordinate q on B in a neighborhood of a point where the manifold degenerates and around which the monodromy is maximally unipotent. The Yukawa couplings, extended as in (3.21) are then simply described by a function

(3.38) $$\Phi(q) = Y_2^{\kappa'}\left(q\frac{\partial}{\partial q}, q\frac{\partial}{\partial q}, q\frac{\partial}{\partial q}\right).$$

3. The Candelas–de la Ossa–Green–Parkes calculation

Consider the quintic hypersurfaces X in $\mathbb{P}^4$ defined by a homogeneous equation F of degree 5. These are Calabi–Yau threefolds satisfying

(3.39) $$h^{2,1}(X) = 101, \quad h^{1,1}(X) = 1.$$

The particular case of the Fermat manifold X_F
$$F = \sum_{i=0}^{4} X_i^5$$
was mentioned in 4.1, where it was "shown" that the mirror of X_F was the desingularization (Theorem 2.6) of the quotient of X_F by the group $G \subset (\mathbb{Z}/5\mathbb{Z})^5/\widetilde{\operatorname{diag}}$ defined by the condition $\sum_i \alpha_i = 0$. The manifold $Y = \widetilde{X_F/G}$ is thus a Calabi–Yau manifold and satisfies

(3.40) $$h^{2,1}(Y) = 1 = h^{1,1}(T_Y), \quad h^{,1}(Y) = 101.$$

The family of dimension 1 of deformations of Y is easy to find. For every $\lambda \in \mathbb{C}$ the variety X_λ having equation
$$F_\lambda = \sum_{i=0}^{4} X_i^5 + \lambda \prod_{i=0}^{4} X_i$$

is also invariant under G, and the deformations Y_λ of Y are given by the quotients $\widetilde{X_\lambda/G}$. In fact, the correct parameter is actually $\psi = \lambda^5$, for X_λ is isomorphic to $X_{\eta\lambda}$ when $\eta^5 = 1$.

When λ tends to infinity in $\mathbb{P}^1$, the manifold X_λ degenerates. It is proved in [43] that the monodromy is maximally unipotent and satisfies the condition (∗). We thus have a canonical coordinate q at infinity and a trivialization κ' of $\overline{\mathcal{H}}_0^{3,0}$ defined up to sign.

We identify $H^2(X, \mathbb{C})/2i\pi H^2(X, \mathbb{Z})$ with $\mathbb{C}/2i\pi\mathbb{Z}$ using the generator $H = c_1(\mathcal{O}_X(1))$ of $H^2(X, \mathbb{Z})$. The mirror map (see 8.1) should furnish a map $M : U \to \mathbb{P}^1$, where the open set

(3.41) $$U \subset H^2(X, \mathbb{C})/2i\pi H^2(X, \mathbb{Z})$$

is defined by the condition $\operatorname{Re} z > 0$.

Let t be the parameter on the universal covering of U defined by

$$\omega_t = -2i\pi t H.$$

Then t satisfies $\operatorname{Im} t > 0$. Let $t' = \dfrac{1}{2i\pi} \log q$ be the canonical parameter at infinity constructed above (and defined up to an integer constant). Then for $|q| < 1$ we should have $\operatorname{Im} t' > 0$.

The first assumption of [43] is that the mirror map is described at infinity by the condition $M^*(t') = t$.

On the other hand (see the Introduction or Sections 5 and 6 of Chapter 2), M_* should be compatible with the Yukawa couplings Y_1, Y_2 when Y_2 is correctly normalized, in the sense that

$$Y_1(\omega)(\phi, \phi, \phi) = Y_2(M(\omega))(M_*\phi, M_*\phi, M_*\phi).$$

The second assumption is that the trivialization of $\overline{\mathcal{H}}_0^{3,0}$ defined in (3.37) gives up to sign the correct normalization of Y_2. The identification of the Yukawa couplings Y_1 and Y_2 then provides

(3.42) $$Y_2^{\kappa'}(e^{2i\pi t})\Big(\frac{1}{2i\pi}\frac{\partial}{\partial t}, \frac{1}{2i\pi}\frac{\partial}{\partial t}, \frac{1}{2i\pi}\frac{\partial}{\partial t}\Big) = Y_1(-2i\pi tH)(H, H, H).$$

In other words

$$\Phi(q) = Y_1(-2i\pi t H)(H, H, H),$$

where $q = e^{2i\pi t}$ and Φ is the holomorphic function defined in (3.38). On the other hand, when we take account of the calculation of $n(\alpha)$ (see Section 6 of Chapter 5), formula (2.54) provides the formula

(3.43) $$Y_1(e^{-2i\pi tH})(H, H, H) = \int_X H^3 + \sum_{d>0} n(d) d^3 \frac{e^{2i\pi dt}}{(1 - e^{2i\pi dt})}$$

where $n(d)$ is the number of rational curves of degree d generically embedded in a general quintic X assuming that the curves are rigid. By comparing the expansions of each of these functions in series of powers of $e^{2i\pi t}$, we see that knowing $\Phi(q)$ as a series of integer powers of q makes it possible to predict the number of rational curves on a general quintic in any degree. The predictions have been verified up to degree 4 (see [45], [46], and [51]).[1]

---

[1] They have now been proved by Givental [G].

3.1. The calculation of normalized Yukawa couplings.

It remains to be seen how one can compute the function $\Phi(q)$ of (3.38). Calculating the Yukawa couplings for an algebraic normalization of κ and in algebraic coordinates presents no difficulty (according to [44] and [47], it is a purely algebraic calculation (see 5)). The principal difficulty is to calculate the normalized section κ' of (3.37) and the canonical coordinate q, which has a manifestly transcendant character. In the following section we explain the notion of a Picard–Fuchs equation and calculate these equations more or less explicitly. In § 5 we show how these equations make it possible to conclude the argument of [43].

4. Picard–Fuchs equations

Let B be a smooth curve and $\pi : \mathcal{X} \to B$ a proper smooth map. Let ω be a holomorphic section of the bundle $\mathcal{H}^{n,0} = R^0 \pi_* K_{\mathcal{X}/B}$. Let t be a coordinate on B. By using the Gauss–Manin connection ∇ on $\mathcal{H}^n$ (see 5.1), one can define holomorphic sections ω_k of $\mathcal{H}^n$ by

$$(3.44) \qquad \omega_k = (\nabla_{\partial/\partial t})^k(\omega),$$

For a certain $N \leq \dim \mathcal{H}^n$, we have certainly a relation with meromorphic coefficients

$$(3.45) \qquad \omega_N = \sum_{i<N} \alpha_i(t) \omega_i.$$

DEFINITION 3.11. The equation $\dfrac{\partial^N \phi}{\partial t^N} = \sum_{i<N} \alpha_i(t) \dfrac{\partial^i \phi}{\partial t^i}$ is called the *Picard–Fuchs equation* of $(\mathcal{X}, \omega)$.

Suppose now that $N = \dim \mathcal{H}^n$, that is, that the $\omega_k(b)$ form a basis of $\mathcal{H}^n_b = H^n(X_b, \mathbb{C})$ for b generic and $0 \leq k \leq N-1$. We then have:

LEMMA 3.12. *The solutions of the Picard–Fuchs equation are the (multivalued) functions $\phi_\alpha = \langle \alpha, \omega \rangle$, where the α are flat sections of $\mathcal{H}_n := (\mathcal{H}^n)^*$.*

In fact it is clear that the ϕ_α are solutions, since by the flatness of α, we have

$$\frac{d}{dt}(\langle \alpha, \omega \rangle) = \langle \alpha, \nabla_{\partial/\partial t}(\omega) \rangle.$$

On the other hand, since the $\omega_k(b)$, where $0 \leq k \leq N-1$, form a base of the fiber $\mathcal{H}^n_b$ for b generic, the ϕ_α generate a space of dimension N of solutions and thus generate all the solutions. □

We consider families of dimension 1 of hypersurfaces $X_t \subset \mathbb{P}^{n+1}$. We study the variation of Hodge structure on $H^n(X_t)$, or, as in the case described above, on $H^n(X_t)^G$, where G is a finite group acting on X_t. For simplicity we assume that the family X_t is given by a pencil $F_t = F + tH$, where H is G-invariant in the second case. We shall show how the representation of the cohomology of X_t by residues makes it possible to calculate the Picard–Fuchs equations.

4.1. Calculation of the cohomology of X_t by residues.

Let $X \subset \mathbb{P}^{n+1}$ be a smooth hypersurface having equation $F = 0$ and let $U = \mathbb{P}^{n+1} - X$. We have the residue map

$$H^{k+1}(U) \longrightarrow H^k(X)$$

(obtained by integration on the fiber of the boundary of a tubular neighborhood of X in $\mathbb{P}^{n+1}$), which provides an isomorphism

$$H^{n+1}(U) \cong \mathrm{Ker}\left(H^n(X) \to H^{n+2}(\mathbb{P}^{n+2})\right),$$

as one can see by combining the exact sequence of relative cohomology of the pair $(\mathbb{P}^{n+2}, U)$ and the Thom isomorphism (see [**47**]). If n is odd, we have $H^{n+1}(U) \cong H^n(X)$. Since U is affine, $H^{n+1}(U)$ is the quotient of the space of closed holomorphic $(n+1)$-forms over the space of exact holomorphic forms. Indeed

(3.46) $$0 \to \mathcal{O}_U \to \cdots \to \Omega_U^{n+1} \to 0$$

is a resolution of $\mathbb{C}$ over U, and we have $H^i(\Omega_U^p) = 0$ for $i > 0$. In fact, according to Grothendieck [**49**], one may consider only holomorphic forms on U that are meromorphic along X. These forms are written

$$\frac{P}{F^k}\Omega, \quad \text{where} \quad \Omega = \sum_i (-1)^i X_i \, dX_0 \wedge \cdots \wedge \widehat{dX_i} \wedge \cdots \wedge dX_{n+1}$$

and P is a homogeneous polynomial of degree $(kd - n - 2)$ if $d = \deg F$.

The following lemma is a consequence of Theorem 3.14, which will be proved below.

LEMMA 3.13 (Griffiths [**47**]). *The class of the form* $\dfrac{P}{F^k}\Omega$ *is zero on* $H^{n+1}(U, \mathbb{C})$ *if and only if*

$$\frac{P}{F^k}\Omega = d\Phi, \quad \Phi \in H^0(\Omega^n_{\mathbb{P}^{n+1}}((k-1)X)).$$

We now consider a pencil

$$F_t = F + tH$$

and let

$$\omega_t = \mathrm{Res}_{X_t}(P\Omega/F_t)$$

be a holomorphic section of the bundle $t \to H^0(K_{X_t})$. It is clear that the Gauss–Manin connection for the cohomology of the open sets U_t is obtained by differentiating the forms $P_t\Omega/F_t^k$ with respect to t, and as the residue map is flat with respect to the Gauss–Manin connection associated with the deformations of X and U, the result is:

(3.47) $$\nabla_{\partial/\partial t}\omega_t = \mathrm{Res}_{X_t}\left(\frac{-HP}{F_t^2}\Omega\right).$$

By continuing to differentiate with respect to t and applying Lemma 3.13, we see that to obtain the Picard–Fuchs equation it suffices to find functions $\alpha_i(t)$ for $0 \le i \le N_1$ (which in fact are algebraic) and a form $\Phi \in H^0(\Omega^n_{\mathbb{P}^{n+1}}(NX))$ such that

(3.48) $$d\Phi = (-1)^N(N!)\frac{H^N P}{F^{N+1}}\Omega - \sum_{i<N}\alpha_i(t)(-1)^i(i!)\frac{H^i P}{F^{i+1}}\Omega.$$

4.2. Hodge filtration and pole-order filtration.

The following theorem, which sharpens Lemma 3.13, makes the calculation of the coefficients $\alpha_i(t)$ of the Picard–Fuchs equation very simple, at least for the case of the family X_λ of 3.

THEOREM 3.14 (Griffiths). (i): *The residues along $X = V(F)$ of the meromorphic forms $\dfrac{P}{F^{k+1}}\Omega$ generate $F^{n-k}H^n_{\mathrm{prim}}(X)$.*

Also,

(ii): *The class of* $\mathrm{Res}\left(\dfrac{P}{F^{k+1}}\Omega\right)$ *is zero if and only if* $\dfrac{P}{F^{k+1}}\Omega = d\Phi$ *with* $\Phi \in H^0(\Omega^n_{\mathbb{P}^{n+1}}(kX))$.

(iii): *A necessary and sufficient condition for the existence of a polynomial Q and a form $\Phi \in H^0(\Omega^n_{\mathbb{P}^{n+1}}(kX))$ such that*

$$\frac{P}{F^{k+1}}\Omega = \frac{Q}{F^k}\Omega + d\Phi,$$

is that P belong to $J_F^{(k+1)d-n-2}$, where J_F is the Jacobian ideal generated by the partial derivatives $\partial F/\partial X_i$.

The proof of (i) and (ii) proceeds as follows.

Denoting the sheaf of closed holomorphic differential forms of degree $\bullet$ on U that are meromorphic with a pole of order at most k along X by ${}^c\Omega^\bullet_{\mathbb{P}^{n+1}}(kX)$, we have for $k \geq 2$ and $p+1 \geq 2$ an exact sequence

$$(3.49) \qquad 0 \to {}^c\Omega^p_{\mathbb{P}^{n+1}}\big((k-1)X\big) \to \Omega^p_{\mathbb{P}^{n+1}}\big((k-1)X\big) \xrightarrow{d} {}^c\Omega^{p+1}_{\mathbb{P}^{n+1}}(kX) \to 0.$$

(Indeed, one can verify that a closed form of degree at least 2 with a pole of order k along X is locally, in a neighborhood of X, the differential of a form having a pole of order $(k-1)$ along X.) Since the holomorphic forms of degree $n+1$ on $\mathbb{P}^{n+1}$ are closed, we have

$${}^c\Omega^{p+1}_{\mathbb{P}^{n+1}}\big((k+1)X\big) = \Omega^{n+1}_{\mathbb{P}^{n+1}}\big((k+1)X\big),$$

and we thus obtain for $k \leq n$ a series of short exact sequences:

$$(3.50) \quad \begin{cases} 0 \to {}^c\Omega^n_{\mathbb{P}^{n+1}}(kX) \to \Omega^n_{\mathbb{P}^{n+1}}(kX) \xrightarrow{d} \Omega^{n+1}_{\mathbb{P}^{n+1}}\big((k+1)X\big) \to 0, \\ 0 \to {}^c\Omega^{n-1}_{\mathbb{P}^{n+1}}\big((k-1)X\big) \to \Omega^{n-1}_{\mathbb{P}^{n+1}}\big((k-1)X\big) \xrightarrow{d} {}^c\Omega^n_{\mathbb{P}^{n+1}}(kX) \to 0, \\ \cdots \\ 0 \to {}^c\Omega^{n-k+1}_{\mathbb{P}^{n+1}}(X) \to \Omega^{n-k+1}_{\mathbb{P}^{n+1}}(X) \xrightarrow{d} {}^c\Omega^{n-k+2}_{\mathbb{P}^{n+1}}(2X) \to 0. \end{cases}$$

Finally, we have the exact sequence

$$(3.51) \qquad 0 \to {}^c\Omega^{n-k+1}_{\mathbb{P}^{n+1}} \to {}^c\Omega^{n-k+1}_{\mathbb{P}^{n+1}}(X) \xrightarrow{\mathrm{Res}} {}^c\Omega^{n-k}_X \to 0.$$

But it follows from Hodge theory that $F^{n-k}H^n(X) \cong H^k({}^c\Omega^{n-k}_X)$. Applying the Bott vanishing theorem to the long exact sequences associated with (3.50), we find a series of isomorphisms

$$(3.52) \quad \begin{cases} H^0\big(\Omega^{n+1}_{\mathbb{P}^{n+1}}((k+1)X)\big)/d\big(H^0(\Omega^n_{\mathbb{P}^{n+1}}(kX))\big) \cong H^1\big({}^c\Omega^n_{\mathbb{P}^{n+1}}(kX)\big), \\ H^1\big({}^c\Omega^n_{\mathbb{P}^{n+1}}(kX)\big) \cong H^2\big({}^c\Omega^{n-1}_{\mathbb{P}^{n+1}}((k-1)X)\big). \\ \cdots \\ H^{k-1}\big({}^c\Omega^{n-k+2}_{\mathbb{P}^{n+1}}(2X)\big) \cong H^k\big({}^c\Omega^{n-k+1}_{\mathbb{P}^{n+1}}(X)\big). \end{cases}$$

Finally (3.51) provides an isomorphism

(3.53)
$$H^k\big({}^c\Omega^{n-k+1}_{\mathbb{P}^{n+1}}(X)\big) \cong \operatorname{Ker}\big\{H^k\big({}^c\Omega^{n-k}_X\big) \to H^{k+1}\big({}^c\Omega^{n-k+1}_{\mathbb{P}^{n+1}}\big)\big\} = F^{n-k}H^n(X)_{\text{prim}}.$$

We have thus shown that
$$H^0\big(\Omega^{n+1}_{\mathbb{P}^{n+1}}((k+1)X)\big)/dH^0\big(\Omega^n_{\mathbb{P}^{n+1}}(kX)\big)$$
is isomorphic to $F^{n-k}H^n_{\text{prim}}(X)$, which proves (i) and (ii).

To prove (iii), we note that the sections of $\Omega^n_{\mathbb{P}^{n+1}}(kX)$ are of the form
$$\Phi = \sum_i \frac{P_i}{F^k}\Omega_i$$
where $\Omega_i = \operatorname{int}(\partial/\partial X_i)(\Omega)$ and $\deg P_i = kd - n - 1$. We thus have

(3.54)
$$\begin{aligned}d\Phi &= \frac{d\big(\sum_i P_i \Omega_i\big)}{F^k} - k\frac{dF}{F^{k+1}} \wedge \sum_i P_i\Omega_i \\ &= \frac{d\big(\sum_i P_i\Omega_i\big)}{F^k} - k\frac{\big(\sum_i P_i \partial F/\partial X_i\big)}{F^{k+1}}\Omega - \deg F \frac{A}{F^k},\end{aligned}$$

where
$$A = \sum_i (-1)^i P_i\, dX_0 \wedge \cdots \wedge \widehat{dX_i} \wedge \cdots \wedge dX_{n+1}.$$

By identifying the terms of maximal degree in $1/F$, we conclude from this that
$$\frac{P}{F^{k+1}}\Omega = d\Phi + \frac{Q}{F^k}\Omega$$
for $\Phi \in H^0\big(\Omega^n_{\mathbb{P}^{n+1}}(kX)\big)$ as above, is equivalent to
$$P = -k\sum_k P_i \frac{\partial F}{\partial X_i},$$
which proves (iii). $\square$

Now consider the family X_λ of § 3 and the form $\omega_\lambda = \operatorname{Res}_{X_\lambda}\Omega/F_\lambda$, where:
- $F_\lambda = F + \lambda H$,
- F is the Fermat equation,
- H is the G-invariant polynomial $\prod_i X_i$.

The form ω_λ is G-invariant. The residues $\operatorname{Res}_{X_\lambda}(H^k/F_\lambda^{k+1})\Omega$ are thus elements of $H^3(X_\lambda)^G \cong H^3(Y_\lambda)$. It is known by § 4.1 that
$$(\nabla_{\partial/\partial\lambda})^i(\omega) = (-1)^i i!\operatorname{Res}_{X_\lambda}\frac{H^i}{F_\lambda^{i+1}}\Omega.$$

For $i \leq 3$, one can verify that the image of $(-1)^i i!\operatorname{Res}_{X_\lambda}\dfrac{H^i}{F_\lambda^{i+1}}\Omega$ under the projection $F^{3-i}H^3(Y_\lambda) \to H^{3-i,i}(Y_\lambda)$ is nonzero (the target space is of rank one for all i).

The calculation of the coefficients $\alpha_i(\lambda)$ of the Picard–Fuchs equation for $0 \leq i \leq 3$ is thus carried out in the following manner. It is known that the Jacobian ideal J_{F_λ} is equal to the ring of polynomials in degrees strictly larger than 15. We can thus write:

$$(-1)^4 4! H^4 = \sum_i P_i \frac{\partial F}{\partial X_i}. \tag{3.55}$$

By formula (3.54) the P_i make it possible to construct explicitly a form $\Phi \in H^0(\Omega^3_{\mathbb{P}^4}(3X))$ and a polynomial Q_4 of degree 15, both G-invariant, such that:

$$(-1)^4 4! \frac{H^4}{F_\lambda^5} \Omega = d(\Phi) + \frac{Q_4}{F_\lambda^4} \Omega. \tag{3.56}$$

By Lemma 3.13 we therefore have

$$(\nabla_{\partial/\partial\lambda})^4(\omega) = (-1)^4 4! \operatorname{Res}_{X_\lambda} \frac{H^4}{F_\lambda^5} \Omega = \operatorname{Res}_{X_\lambda} \frac{Q_4}{F_\lambda^4} \Omega. \tag{3.57}$$

The coefficient $\alpha_3(\lambda)$ of the Picard–Fuchs equation is then determined as follows. We should have

$$(\nabla_{\partial/\partial\lambda})^4(\omega) = \sum_{i=0}^{3} \alpha_i(\lambda) (\nabla_{\partial/\partial\lambda})^i(\omega). \tag{3.58}$$

But the $(\nabla_{\partial/\partial\lambda})^i(\omega)$ for $i < 3$ generate $F^2 H^3(Y_\lambda)$. Thus $\alpha_3(\lambda)$ is determined by the condition that $\operatorname{Res}_{X_\lambda} Q_4/F_\lambda^4 \Omega$ and $\alpha_3(\lambda) \operatorname{Res}_{X_\lambda} (-1)^3 3! H^3/F_\lambda^4 \Omega$ have the same projection on $H^{0,3}(Y_\lambda)$, which by Theorem 3.14 (iii) is equivalent to

$$Q_4 - \alpha_3(\lambda)(-1)^3 3! H^3 \in J_{F_\lambda}^{15}. \tag{3.59}$$

The other coefficients can be calculated in the same way by successive reduction of the order of the pole.

REMARK 3.15. We have calculated the Picard–Fuchs equation of the family Y_λ endowed with the G-invariant form ω_λ. It is actually preferable to work with the form $\omega'_\lambda = \lambda \omega_\lambda$, which provides a holomorphic section of the bundle $\mathcal{H}^{3,0}(Y_\lambda)$ invariant under the transformations $\lambda \mapsto \zeta \lambda$ with $\zeta^5 = 1$ and nonzero at infinity. The Picard–Fuchs equation of ω'_λ can be derived immediately from that of ω_λ by linear transformations $(\nabla_{\partial/\partial\lambda} \omega'_\lambda = \lambda \nabla_{\partial/\partial\lambda} \omega_\lambda + \omega_\lambda \cdots)$. Finally, to obtain the Picard–Fuchs equation of (Y_ψ, ω'_ψ), $\psi = \lambda^5$, it suffices to note that

$$\frac{\partial}{\partial \psi} = \frac{1}{5\lambda^4} \frac{\partial}{\partial \lambda}$$

and to carry out the obvious linear transformations.

5. Conclusion of the argument

We first explain how the Yukawa couplings are computed for the family X_λ normalized by the section ω, that is, the function $Y_2^\omega(\partial/\partial\lambda, \partial/\partial\lambda, \partial/\partial\lambda)$ with parameter λ, where $\omega = \operatorname{Res}_{X_\lambda} \Omega/F_\lambda$. To do this we use Lemma 3.2 and its proof, which yield

$$Y_2^\omega\left(\frac{\partial}{\partial\lambda}, \frac{\partial}{\partial\lambda}, \frac{\partial}{\partial\lambda}\right) = \left\langle \omega, \left((\nabla_{\partial/\partial\lambda})^3 \omega\right)^{0,3} \right\rangle, \tag{3.60}$$

where "0,3" denotes projection on $\mathcal{H}^{0,3}$ and the bracket is the Serre duality between $H^0(K_{X_\lambda})$ and $H^3(\mathcal{O}_{X_\lambda})$. We have seen that

$$(\nabla_{\partial/\partial\lambda})^3(\omega) = (-1)^3 3! \operatorname{Res}_{X_\lambda} \frac{H^3}{F_\lambda^4}\Omega.$$

Finally, by Theorem 3.14, we have a surjective map

(3.61)
$$H^0\bigl(\mathcal{O}_{\mathbb{P}^4}(15)\bigr)^G \to H^3(X_\lambda, \mathbb{C})^G \cong H^3(Y_\lambda, \mathbb{C}),$$
$$P \mapsto \operatorname{Res}_{X_\lambda} P\Omega/F_\lambda^4,$$

which induces an isomorphism

(3.62)
$$H^0\bigl(\mathcal{O}_{\mathbb{P}^4}(15)\bigr)^G / J_{F_\lambda}^{15} \cong H^3(\mathcal{O}_{Y_\lambda}).$$

To calculate the right-hand side of (3.60), it thus remains only to use the following theorem, due to Griffiths and Carlson [**44**]. Let F be a homogeneous polynomial of degree d on $\mathbb{P}^{n+1}$ and R_F its Jacobian ring, that is, the quotient of the ring of polynomials of $\mathbb{P}^{n+1}$ by the Jacobian ideal of F.

THEOREM 3.16. *If F defines a smooth hypersurface $X \subset \mathbb{P}^{n+1}$, there exists an isomorphism* $\eta : R_F^{(n+2)d-2(n+2)} \cong \mathbb{C}$, *such that the isomorphisms* $H^k(X, \Omega_X^{n-k})_{\text{prim}} \cong R_F^{(k+1)d-n-2}$ *provided by Theorem 3.14 identify the cup product*

(3.63)
$$H^k(X, \Omega_X^{n-k})_{\text{prim}} \otimes H^{n-k}(X, \Omega_X^k)_{\text{prim}} \to \mathbb{C}$$

with the product

(3.64)
$$R_F^{(k+1)d-n-2} \otimes R_F^{(n-k+1)d-n-2} \to R_F^{(n+2)d-2(n+2)}$$

followed by the isomorphism η.

Explicitly, η is simply induced up to a universal coefficient by the map

$$P \mapsto \operatorname{Res}_0\Bigl(\frac{P}{\partial F/X_0 \cdots \partial F/\partial X_{n+1}} dX_0 \wedge \cdots \wedge dX_{n+1}\Bigr).$$

REMARK 3.17. To calculate the Yukawa couplings of the family Y_λ normalized by the section ω (which can now be seen as a section of the bundle having fiber $H^{3,0}(Y_\lambda)$), mapped to the field $\partial/\partial\lambda$, it suffices to calculate them for the family X_λ and divide by the cardinality of G. On the other hand,

$$Y_2^{\omega_\psi'}\Bigl(\frac{\partial}{\partial\psi}, \frac{\partial}{\partial\psi}, \frac{\partial}{\partial\psi}\Bigr) = \Bigl(\frac{1}{5^3}\lambda^{10}\Bigr) Y_2^{\omega_\lambda}\Bigl(\frac{\partial}{\partial\lambda}, \frac{\partial}{\partial\lambda}, \frac{\partial}{\partial\lambda}\Bigr).$$

Assume that we have made the change of coordinates $q = q(1/\psi)$ at infinity and that we have replaced ω_ψ' with $\kappa_\psi' = f(\psi)\omega_\psi'$. We then have

(3.65)
$$\Phi(q) = Y_2^{\kappa'}\Bigl(q\frac{\partial}{\partial q}, q\frac{\partial}{\partial q}, q\frac{\partial}{\partial q}\Bigr) = f(\psi)^2 Y_2^{\omega_\psi'}\Bigl(q\frac{\partial}{\partial q}, q\frac{\partial}{\partial q}, q\frac{\partial}{\partial q}\Bigr)$$
$$= \frac{f(\psi)^2 q(1/\psi)^3 \psi^6}{q'(1/\psi)^3} Y_2^{\omega_\psi'}\Bigl(\frac{\partial}{\partial\psi}, \frac{\partial}{\partial\psi}, \frac{\partial}{\partial\psi}\Bigr)$$

which can be written as a function of q by inverting the series $q(1/\psi)$.

Finally, the Picard–Fuchs equation of the family Y_ψ and the section ω_ψ' are known from 4. The unknown function $f(\psi)$ is equal to $1/\int_{e_0} \omega_\psi'$, where e_0 generates the invariant integer homology by monodromy around infinity (see (3.33)). But it is known that $\int_{e_0} \omega_\psi'$ generates over $\mathbb{C}$ the solutions of the Picard–Fuchs equation without monodromy around infinity (Lemma 3.12). The function $f(\psi)$ is thus determined by the Picard–Fuchs equation up to a multiplicative coefficient β. Similarly, the function $q(1/\psi)$ is characterized by the formula

$$(3.66) \qquad \frac{1}{2i\pi}\log(1) = \frac{\int_{e_1} \omega'_\psi}{\int_{e_0} \omega'_\psi},$$

where e_1 must be an integer homology class satisfying the condition $Ne_1 = e_0$. If one has determined e_0, $\int_{e_1} \omega'_\psi$ must be a solution of the Picard–Fuchs equation having as monodromy precisely $\int_{e_0} \omega'_\psi$ around infinity and that determines it modulo functions of the type $\alpha \int_{e_0} \omega'_\psi$, with $\alpha \in \mathbb{C}$. If the two complex coefficients β and α above have been fixed, the formula (3.65) and the expansion of the solutions of the Picard–Fuchs equation provide the function $\Phi(q)$ as a series in integer powers of q.

Obviously it yet remains to distinguish the solutions of the Picard–Fuchs equation corresponding to the *integral* homology classes e_0, e_1. Since one need only fix the two constants α and β introduced above, this can be done by making the cycles of integration e_0 and e_1 explicit and calculating the first terms of the asymptotic expansion of $\int_{e_i} \omega'_\psi$.

CHAPTER 4

The work of Batyrev

The purpose of this chapter is to describe the early work of Batyrev on mirror symmetry between hypersurfaces in toric varieties. We explain, following Fulton [66], the construction of a toric variety of dimension n starting from a "fan" in $\mathbb{R}^n$, that is, a decomposition of $\mathbb{R}^n$ into convex rational polyhedral cones.

We develop the translation of certain geometric properties or notions (smoothness or $\mathbb{Q}$-factoriality, Cartier divisors, canonical divisor, amplitude of invertible bundles) in combinatorial terms (simplicial cones on $\mathbb{Z}$ or $\mathbb{Q}$, piecewise linear functions on $\mathbb{R}^{n+1}$, support function, and convexity).

This makes it possible to establish easily the bijective correspondence beween toric Fano varieties (the canonical divisor must be a Cartier divisor and have ample inverse) and reflexive polyhedra. Batyrev constructs the mirror symmetry $\{X\} \mapsto \{X'\}$ between certain desingularizations of hypersurfaces having trivial canonical bundle in toric Fano varieties as the involution that associates with a reflexive polyhedron its dual.

In a second step we explain how one can calculate (in principle) the $h^{p,q}$ of certain desingularizations of these hypersurfaces, using the results of Danilov and Khovanskii, and, following Batyrev, we show the equality $h^1(T_X) = h^1(\Omega_{X'})$, where X and X' are as above, which is the first of the predictions made by physicists concerning the comparison of the Hodge numbers of X and its mirror.

1. Toric varieties

Let N be a free $\mathbb{Z}$-module of rank n and M its dual. On $N_\mathbb{R}$ we prescribe a system Σ (fan) of convex poyhedral and rational cones σ, that is, cones generated as convex cones by a finite number of points of N such that $\sigma \cap -\sigma = \{0\}$ and satisfying the following conditions:

(i): If σ' is a face of σ and $\sigma \in \Sigma$, then $\sigma' \in \Sigma$.
(ii): If σ and σ' are in Σ, then $\sigma \cap \sigma'$ is a face of σ and a face of σ'.

We then construct a toric variety $\mathbb{P}_\Sigma$ of dimension n associated with Σ as follows. We endow M with a basis, so that $m \in M$ can be written $m = (m_1, \ldots, m_n)$. For all $\sigma \in \Sigma$ we set

$$U_\sigma = \operatorname{Spec} A_\sigma, \quad \text{where} \quad A_\sigma = \mathbb{C}\left[\prod X_i^{m_i}\right]_{(m_i) \in \sigma^*},$$

and the dual σ^* is the set of points $m \in M$ that are positive or zero on σ.

The U_σ form an affine covering of $\mathbb{P}_\Sigma$.

If $\sigma' \subset \sigma$ is a face, we have an inclusion $\sigma^* \hookrightarrow \sigma'^*$, providing a morphism $A_\sigma \to A_{\sigma'}$ which one can easily verify induces an open inclusion $j_{\sigma'\sigma} : U_{\sigma'} \hookrightarrow U_\sigma$.

The open sets U_σ are glued together in $\mathbb{P}_\Sigma$ by using the property (ii). Let $\sigma'' = \sigma \cap \sigma'$; then the intersection $U_\sigma \cap U_{\sigma'}$ in $\mathbb{P}_\Sigma$ is equal to $U_{\sigma''}$, which is naturally contained in U_σ via $j_{\sigma''\sigma}$ and in $U_{\sigma'}$ via $j_{\sigma''\sigma'}$.

1.1. Compactness and smoothness.

LEMMA 4.1. *The variety $\mathbb{P}_\Sigma$ is complete if and only if $N_\mathbb{R}$ is equal to the union of the cones of Σ.*

To see that this condition is necessary, we note that $\mathbb{P}_\Sigma$ contains
$$U_0 = \operatorname{Spec} A_0 = \operatorname{Spec} \mathbb{C}[X_1^\pm, \ldots, X_n^\pm] = (\mathbb{C}^*)^n.$$
Let $n = (n_1, \ldots, n_n) \in N$. Then U_0 contains $(\lambda^{n_1}, \ldots, \lambda^{n_n})$ for $\lambda \in \mathbb{C}^*$. If $\mathbb{P}_\Sigma$ is complete, there should exist a cone σ such that the limit $\lim_{\lambda \to 0}(\lambda^{n_1}, \ldots, \lambda^{n_n})$ exists in U_σ. But clearly this limit exists if and only if $n \in \sigma$ (see [66] for the converse). $\square$

Throughout the following we assume that $\mathbb{P}_\Sigma$ is complete.

LEMMA 4.2. *The variety $\mathbb{P}_\Sigma$ is smooth if and only if every cone $\sigma \in \Sigma$ is generated by $e_1 \ldots, e_{k(\sigma)}$, where the e_i are independent elements of N that can be completed to a $\mathbb{Z}$-basis of N.*

Indeed, the sufficiency of the condition comes from the fact that if σ is generated by $e_1, \ldots, e_{k(\sigma)}$, which one can complete to a basis $e_1, \ldots, e_n$ of N in the dual basis, σ^* is described by the condition $m_i \geq 0$, where $i \leq k(\sigma)$, and we thus have an isomorphism
$$A_\sigma \cong \mathbb{C}[X_1, \ldots, X_{k(\sigma)}, X_{k(\sigma)+1}^\pm, \ldots, X_n^\pm]$$
and hence also
$$U_\sigma \cong \mathbb{C}^{k(\sigma)} \times (\mathbb{C}^*)^{n-k(\sigma)}$$
(see [66] for the converse). $\square$

1.2. The action of $(\mathbb{C}^*)^n$.

The torus $T := (\mathbb{C}^*)^n$ acts on every open set U_σ, that is, on every algebra A_σ by

(4.1) $\quad \lambda = (\lambda_1, \ldots, \lambda_n), \quad \lambda^*(X_1^{m_1} \cdots X_n^{m_n}) = \lambda_1^{m_1} \cdots \lambda_n^{m_n} X_1^{m_1} \cdots X_n^{m_n}.$

These actions are clearly compatible on the intersections $U_\sigma \cap U_{\sigma'}$, and thus provide an action of T on $\mathbb{P}_\Sigma$, justifying the term *toric variety*. The action of T also makes it possible to construct a stratification of $\mathbb{P}_\Sigma$ by the T-orbits, which are isomorphic to $(\mathbb{C}^*)^k$. One can describe this stratification explicitly in terms of the structure of the fan Σ as follows. Let σ be a cone of Σ; let $N_\sigma \subset N$ be the sublattice generated by σ, and let $M_\sigma := N_\sigma^\perp$. We then define $O_\sigma \subset U_\sigma$ by the equations $X_1^{m_1} \cdots X_n^{m_n}$ for $m = (m_1, \ldots, m_n) \in \sigma^* - M_\sigma$. One can verify easily that O_σ is an orbit under T and is isomorphic to $(\mathbb{C}^*)^{n-k(\sigma)}$, where $k(\sigma)$ is the rank of the lattice generated by σ.

2. Weil and Cartier divisors

The *Weil divisors* of an irreducible algebraic variety are defined as combinations of irreducible hypersurfaces with integer coefficients, while the *Cartier divisors* are defined as assignments of meromorphic functions g_i on the open sets U_i of a covering, defined up to an invertible function, and satisfying the condition that g_i/g_j is invertible over $U_i \cap U_j$.

A Cartier divisor has associated with it a Weil divisor (its intersection with the open set U_i is the divisor of the function g_i), but equivalence between the two notions holds only if the variety is normal and locally factorial (the ideals of codimension 1 in the local rings at closed points must be principal ideals).

Since toric varieties do not in general satisfy this last condition, we describe their Weil and Cartier divisors separately below, following [**66**]. We are actually interested in classes of divisors modulo principal divisors, that is, divisors of a meromorphic function on the variety under consideration, and that is why it suffices in both cases to study divisors that are invariant under the action of the torus T.

2.1. Weil divisors. The description of T-invariant Weil divisors is immediate. Indeed, the cone $\{0\} \in \Sigma$ provides a T-orbit $O_0 = U_0$ of dimension n. The T-invariant Weil divisors are thus necessarily the components of $\mathbb{P}_\Sigma - U_0$ of dimension $(n-1)$, which are the closures of the T-orbits of dimension $(n-1)$, that is, orbits O_σ, where σ is a cone of dimension 1. Such a σ possesses a unique generator $v \in N$ (v is the unique integer point not divisible by σ) and we denote the corresponding Weil divisor by D_v.

2.2. Cartier divisors. A T-invariant Cartier divisor gives on each U_σ a rational function ϕ_σ preserved up to a coefficient by the action of T and defined modulo an invertible function on U_σ. On $U_\sigma \cap U_{\sigma'}$ the function $\phi_\sigma/\phi_{\sigma'}$ must be invertible.

The function ϕ_σ, being an eigenvector for the action of T, can necessarily be written as a monomial

$$\phi_\sigma = \prod_i X_i^{m_i(\sigma)}.$$

The fact that this function is defined on U_σ modulo an invertible (T-invariant) function implies that $m(\sigma) := (m_1(\sigma), \ldots, m_n(\sigma))$ is defined modulo M_σ, which shows that to give the function $h_\sigma = \langle m(\sigma), \cdot \rangle$ on $\sigma \subset N_\mathbb{R}$ is an assignment equivalent to giving the restriction of the Cartier divisor in question to U_σ. Finally, the condition of invertibility of $\phi_\sigma/\phi_{\sigma'}$ on $U_\sigma \cap U_{\sigma'} = U_{\sigma''}$ is equivalent to the condition $m(\sigma) - m(\sigma') \in M_{\sigma''}$, or to $h_\sigma = h_{\sigma'}$ on $\sigma \cap \sigma' = \sigma''$. We have thus shown the following.

PROPOSITION 4.3. *There exists a one-o-one correspondence between T-invariant Cartier divisors on $\mathbb{P}_\Sigma$ and functions h on $N_\mathbb{R}$ having the following property: on each cone σ the function h is of the form $\langle m(\sigma), \cdot \rangle$ for a point $m(\sigma) \in M$.*

REMARK 4.4. It is clear from the preceding description that the T-invariant principal Cartier divisors correspond to functions h that are globally defined by an element of M.

It now remains to see how one can pass from Cartier divisors to Weil divisors in the description given above. Let h be a function on $N_\mathbb{R}$ as above, and let D_h be the corresponding Cartier divisor.

LEMMA 4.5. *The associated Weil divisor* $\operatorname{div}(D_h)$ *is equal to* $\sum_v h(v)D_v$, *where v ranges over the set of generators of cones in Σ of dimension 1.*

Indeed, the open sets $\operatorname{Spec} A_\sigma$ for $\dim \sigma = 1$ cover the complement of a set of codimension 2 in $\mathbb{P}_\Sigma$, so that it suffices to calculate $\operatorname{div}(D_h)$ on the corresponding open sets U_σ. Now let $v \in N$ be the integral generator of σ and let $m(\sigma) \in M$ be a point defining $h_{|\sigma}$. By definition D_h is then represented in U_σ by the rational function $X_1^{m_1(\sigma)} \ldots, X_n^{m_n(\sigma)}$. But we recall (see 1.2) that D_v is defined on U_σ by any equation $\prod_i X_i^{m_i}$ such that $\langle m, v \rangle = 1$ and $m = (m_1, \ldots, m_n)$. (If $m(v) = 0$, then $\prod_i X_i^{m_i}$ is invertible in U_σ.) It then follows immediately that the multiplicity of D_v in $\operatorname{div}(D_h) \cap U_\sigma$ is equal to $\langle m(\sigma), v \rangle = h(v)$.

2.3. The canonical divisor. The variety $\mathbb{P}_\Sigma$ is smooth in codimension 1 since the open sets U_σ such that $\dim \sigma = 1$, which cover the complement U of an algebraic subset of codimension 2 according to 1.2, are smooth by virtue of Lemma 4.2. We shall calculate a canonical T-invariant divisor of $\mathbb{P}_\Sigma$, that is, the divisor of a canonical meromorphic form on the open set of smoothness U of $\mathbb{P}_\Sigma$. For that we consider the T-invariant differential form of degree n on $U_0 \cong (\mathbb{C}^*)^n$:

$$(4.2) \qquad \Omega = \frac{dX_1}{X_1} \wedge \cdots \wedge \frac{dX_n}{X_n}.$$

To calculate the pole of this form along the divisor D_v in the open set U_σ (where v generates σ), we take $m \in M$ such that $\langle m, v \rangle = 1$. The algebra A_σ is then equal to $\mathbb{C}[X^m, X^{\pm m'}]$, where $X^m = \prod_i X_i^{m_i}$ and m' ranges over $v^\perp$. For a constant $C \in \mathbb{C}^*$ we then have

$$(4.3) \qquad \Omega = C \frac{dX^m}{X^m} \wedge \frac{dX^{m'_1}}{X^{m'_1}} \wedge \cdots \wedge \frac{dX^{m'_{n-1}}}{X^{m'_{n-1}}},$$

where $\{m'_1, \ldots, m'_{n-1}\}$ is a basis of $v^\perp$. Since the $X^{m'}$ are invertible in U_σ and together with X^m provide coordinates on U_σ, it is clear by (4.3) that Ω does not vanish on U_σ and has exactly one pole of order 1 along D_v.

We have thus exhibited a Weil divisor $\sum_v D_v$ whose restriction to U is anti-canonical. Proposition 4.3 and Lemma 4.5 yield the following proposition immediately.

PROPOSITION 4.6. *The anticanonical divisor of U extends to a Cartier divisor on $\mathbb{P}_\Sigma$ if and only if there exists a function h on $N_\mathbb{R}$ that assumes the value 1 on the generators of cones of Σ of dimension 1 and such that for each $\sigma \in \Sigma$ there exists $m(\sigma) \in M$ for which $h_{|\sigma} = \langle m(\sigma), \cdot \rangle$.*

Such a function h is then called a *support function* of Σ. If it exists, we certainly have

$$(4.4) \qquad K_{\mathbb{P}_\Sigma}^{-1} \cong \mathcal{O}(D_h).$$

3. Polyhedra and toric varieties

Let $\mathbb{P}_\Sigma$ be a toric variety, and let h be a function on $N_\mathbb{R}$ defining a Cartier divisor D_h, that is, given by a collection $\{m(\sigma)\}$ of points of M such that $h = \langle m(\sigma), \cdot \rangle$ on σ. The corresponding Cartier divisor is then given by the collection of functions $X^{m(\sigma)}$ on U_σ. We use the notation

$$L_h = \mathcal{O}(D_h)$$

for the associated invertible sheaf.

3.1. Sections of L_h. Since L_h is T-invariant, its sections are generated by eigenvectors of the action of T, that is, by monomials, if we identify them with functions using the meromorphic section $(X^{m(\sigma)})$ with divisor D_h. A monomial $X^m = \prod_i X_i^{m_i}$ is a section of L_h if and only if on each open set U_σ the section $X^m \cdot X^{m(\sigma)}$ has no poles in U_σ. But, by definition of the algebra A_σ of functions on U_σ we have $X^m \cdot X^{m(\sigma)} \in A_\sigma$ if and only if $m + m(\sigma) \in \sigma^*$. We have thus shown the following:

LEMMA 4.7. *The T-invariant sections of L_h can be identified with the integral points $m \in \Delta_h := \bigcap_\sigma (-m(\sigma) + \sigma^*)$.*

REMARK 4.8. It is clear that distinct integral points of Δ_h provide sections corresponding to different characters of T. Thus the set of integral points of Δ_h is in one-to-one correspondence with a basis of $H^0(L_h)$.

3.2. Global generation and the amplitude of L_h.

LEMMA 4.9. *The divisor L_h is generated by its global sections if and only if the function h is convex.*

The global sections of L_h were described in the preceding lemma.

Now let σ be a cone of Σ of dimension n, and let $p_\sigma \in \mathbb{P}_\Sigma$ be the T-invariant point that corresponds to it (see 1.2). By definition p_σ is described in U_σ by the equations $\prod_i X_i^{m_i} = 0$ for $m = (m_i) \in \sigma^* - \{0\}$. Then let s be a T-invariant section of L_h, and write

$$\frac{s}{X^{m(\sigma)}} = \prod_i X_i^{m_i} \quad \text{where} \quad m \in \Delta_h.$$

Then $\prod_i X_i^{m_i + m_i(\sigma)}$ belongs to A_σ and vanishes at p_σ if and only if s vanishes at p_σ. But this function vanishes at p_σ if $m_i + m_i(\sigma) \neq 0$ for at least one i. The point p_σ is thus not a base point of $|L_h|$ if and only if the monomial $1/\prod_i X_i^{m_i(\sigma)}$ corresponds to a global section of L_h, which is equivalent to

(4.5) $$-m(\sigma) \in \bigcap_{\sigma'} -m(\sigma') + \sigma'^*.$$

This is equivalent to the fact that $\langle m(\sigma), \cdot \rangle$ is less than or equal to $\langle m(\sigma'), \cdot \rangle = h_{|\sigma'}$ on σ'; and since $\langle m(\sigma), \cdot \rangle = h_{|\sigma}$ on σ, this means that the graph of h is supported over its hyperplane of support at a point of σ. Thus h is convex if and only if none of the points p_σ is a base point of $|L_h|$. To complete the proof of the lemma, it suffices to note that the base locus of $|L_h|$ is T-invariant, and hence contains a point p_σ if it is not empty. □

One can show similarly the following result.

LEMMA 4.10. *The invertible bundle L_h is ample if and only if h is strictly pseudoconvex, in the sense that for any cone σ of dimension n and any cone $\sigma' \neq \sigma$ of dimension n the function $\langle m(\sigma), \cdot \rangle$ is less than or equal to h on σ', with strict inequality on $\sigma' - \sigma \cap \sigma'$.*

3.3. Polyhedra and toric varieties. Let D_h be an ample T-invariant Cartier divisor on $\mathbb{P}_\Sigma$. According to Lemma 4.7 the T-invariant sections of L_h^k are in one-to-one correspondence with the integral points of $k\Delta_h$, where the polyhedron Δ_h is defined in Lemma 4.7. Since L_h is ample, we have

$$(4.6) \qquad \mathbb{P}_\Sigma \cong \mathrm{Proj}\Big(\bigoplus_k H^0(\mathbb{P}_\Sigma, L_h^k)\Big) \cong \mathrm{Proj}\Big(\mathbb{C}[X_0^{m_0}\cdots X_n^{m_n}]_{(m_i/m_0)\in\Delta_h}\Big),$$

where the grading of the second ring is given by m_0. If h is strictly positive except at 0 (which is equivalent to $D_h = \sum_v n(v)D_v$, with $n(v) > 0$ for every generator v of a cone of dimension 1 in Σ), it is clear that $\Delta := \Delta_h$ contains 0 in its interior and has integral vertices. Indeed, by definition Δ can then be identified with the dual of the convex polyhedron Δ^* defined by

$$(4.7) \qquad \Delta^* = \{x \in N_\mathbb{R};\ h(x) \leq 1\}.$$

We thus have

$$(4.8) \qquad \Delta = \{x \in M_\mathbb{R};\ \langle x, y\rangle \geq -1,\ \forall y \in \Delta^*\}.$$

The vertices of Δ are thus in one-to-one correspondence with the faces of Δ^* of dimension $(n-1)$, or with cones σ of dimension n in Σ, by the strict pseudoconvexity of h. But by hypothesis the faces of Δ^* of dimension $(n-1)$ are defined by equations $\langle m, \cdot\rangle = 1$ for an integer point $m \in M$. This clearly implies that Δ has integral vertices. On the other hand, by (4.8), 0 lies in the interior of Δ.

Given a convex polyhedron Δ in M satisfying the conditions above, we associate with it, as in (4.6), a projective variety

$$\mathbb{P}_\Delta = \mathrm{Proj}\big(\mathbb{C}[X_0^{m_0}\cdots X_n^{m_n}]_{(m_i/m_0)\in\Delta}\big).$$

We now note the description of $\mathbb{P}_\Delta$ as a polarized toric variety (that is, endowed with an ample T-invariant Cartier divisor). Let $\Delta^* \subset N_\mathbb{R}$ be the dual polyhedron to Δ:

$$(4.9) \qquad \Delta^* = \{x \in N_\mathbb{R};\ \langle x, y\rangle \geq -1,\ \forall y \in \Delta\}.$$

As a system of cones Σ we take the cones on the faces of Δ^*. The polarization, that is, the assignment of an ample Cartier divisor, is given as follows.

We take the function h_Δ, linear on each cone of Σ and such that $h_\Delta(x) \leq 1$ defines the polyhedron Δ^*. We would like to show that h_Δ is defined on every cone σ by an element $m(\sigma) \in M$. It suffices to see this for cones of dimension n, for which it is a consequence of the fact that Δ^* is the dual of Δ and Δ has integer vertices. The cones of dimension n in σ are the cones on the faces of dimension $(n-1)$ of Δ^* defined by an equation $\langle m, \cdot\rangle = 1$, where $-m$ is a vertex of Δ. Hence $m \in M$. On such a cone the function $h = \langle m, \cdot\rangle$ is thus indeed defined by an element of M, which provides a T-invariant Cartier divisor by Proposition 4.3, which is ample by Lemma 4.10.

4. Toric Fano varieties

DEFINITION 4.11. The toric variety $\mathbb{P}_\Sigma$ is a *Fano* variety if its anticanonical divisor is a Cartier divisor and is ample.

By Proposition 4.6 there then exists a support function h_Δ assuming the value 1 on the integer generators of cones of dimension 1 in Σ and defined on every cone σ by an element $m(\sigma) \in M$. According to Lemma 4.10, the amplitude of $K_{\mathbb{P}_\Sigma}^{-1}$ is equivalent to the fact that h is strictly pseudoconvex.

We now give the following characterization of the convex polyhedra Δ with integer vertices containing 0 in their interior, such that the natural polarization of $\mathbb{P}_\Delta$ is equal to the anticanonical Cartier divisor given as the locus of the poles of the section Ω of $K_{\mathbb{P}_\Delta}$ see (4.2)).

LEMMA 4.12. *The natural Cartier divisor D_{h_Δ} of $\mathbb{P}_\Delta$ is $\operatorname{div}(\Omega)$ if and only if Δ^* has integral vertices.*

Indeed, if equality holds, the function h_Δ that determines the polarization must be the support function, and hence must satisfy $h_\Delta(v) = 1$ for every generator of a cone of dimension 1 in Σ. But since Σ is constructed on Δ^* as in 3.3, these cones of dimension 1 are the cones on the vertices of Δ^*. By definition the function h_Δ equals 1 on the vertices of Δ^*. Such a vertex can be written uniquely as λv with $v \in N$ and $\lambda > 0$. If we also have $h_\Delta(v) = 1$, we must thus have $\lambda = 1$ and Δ^* does indeed have integral vertices. Conversely, it can be verified immediately that if Δ^* has integral vertices, the piecewise linear function that equals 1 on the boundary of Δ^* is the support function of the system of cones constructed on Δ^*. □

We now introduce the following notion, due to Batyrev.

DEFINITION 4.13. A convex rational polyhedron $\Delta \subset M_\mathbb{R}$ is *reflexive* if it has integer vertices and contains 0 in its interior, and its dual $\Delta^* \subset N_\mathbb{R}$ satisfies the same conditions.

By biduality it is clear that if Δ is reflexive, then Δ^* is also reflexive. We have now proved the following.

PROPOSITION 4.14 (see [**60**], [**61**]). *The toric Fano varieties of dimension n polarized by $-\operatorname{div}(\Omega)$, where Ω is the meromorphic form of (4.2), are in one-to-one correspondence with the reflexive polyhedra $\Delta \subset M_\mathbb{R}$ with $\operatorname{rank} M = n$. The map $\Delta \mapsto \Delta^*$ between reflexive polyhedra thereby provides an involution on the set of these varieties.*

The connection with mirror symmetry is the following [**60**]. For every toric Fano variety, the family of hypersurfaces in $|K^{-1}|$ provides a family of varieties with trivial canonical bundle (unfortunately singular, in general). The mirror family would then be simply the family of hypersurfaces with trivial canonical bundle on the toric Fano variety obtained by the action of that involution. Because of the singularities, this construction works well only in dimension 3. The following sections explain how one can construct desingularizations with trivial canonical bundle in this case and detail the computation of the Hodge numbers of the resulting varieties, so as to prove the inversion of Hodge numbers predicted by mirror symmetry (see (1.48)).

5. Desingularization

Let $\mathbb{P}_\Delta$ be a toric Fano variety. We shall construct a partial desingularization $\tau : \mathbb{P}_\Sigma \to \mathbb{P}_\Delta$ satisfying the property

(4.10) $$\tau^*(K_{\mathbb{P}_\Delta}) = K_{\mathbb{P}_\Sigma}$$

and which is nonsingular in codimension 3.

In the particular case when $n = 4$ the singularities of $\mathbb{P}_\Sigma$ are isolated, and the generic hypersurfaces in the linear system $\left|\tau^*\!\left(K_{\mathbb{P}_\Delta}^{-1}\right)\right| = \left|K_{\mathbb{P}_\Sigma}^{-1}\right|$ do not intersect the singular locus of $\mathbb{P}_\Sigma$. Hence they are smooth according to Bertini and have trivial canonical bundle by the adjunction formula. (It is known that this linear system is generated by the global sections, according to Lemma 4.9.)

We proceed as follows. The toric variety $\mathbb{P}_\Delta$ is associated with the system of cones on the faces of $\Delta^* \subset N_\mathbb{R}$. We shall now subdivide (or triangulate) the boundary of the polyhedron Δ^* by imposing the following conditions:

(i): the vertices of the triangulation are precisely the points of $N \cap \partial \Delta^*$;

(ii): the faces of dimension k of the subdivided polyhedron are the convex hull of $k + 1$ (integral) vectors.

We then take Σ to be the set of cones on the faces of the subdivided polyhedron. We have a natural map $\tau : \mathbb{P}_\Sigma \to \mathbb{P}_\Delta$ defined as follows. If σ is in Σ, there exists a cone σ' on a face of Δ^* such that $\sigma \subset \sigma'$. We thus have an inclusion σ'^* that provides $\tau^* : A_{\sigma'} \to A_\sigma$ and defines a morphism of U_σ into $U_{\sigma'}$.

These morphisms can be pieced together to form a morphism $\tau : \mathbb{P}_\Sigma \to \mathbb{P}_\Delta$ that is birational since it is an isomorphism on the dense open sets $U_0 = \operatorname{Spec} A_0 \cong (\mathbb{C}^*)^n$ of $\mathbb{P}_\Sigma$ and $\mathbb{P}_\Delta$.

LEMMA 4.15. *The map τ satisfies the condition*

$$(4.11) \qquad \tau^*(K_{\mathbb{P}_\Delta}) = K_{\mathbb{P}_\Sigma}.$$

Indeed, this follows from the functoriality of the description of Cartier divisors in terms of functions on $N_\mathbb{R}$ (Proposition 4.3). One can see immediately that if h defines a Cartier divisor D_h on $\mathbb{P}_\Delta$, as in 2.2, the Cartier divisor $\tau^* D_h$ on $\mathbb{P}_\Sigma$ is defined by the same function h.

We recall that a function h corresponds to the anticanonical divisor if and only if it assumes the value 1 on the integral generators of the cones of dimension 1 (Proposition 4.6). But the anticanonical divisor of $\mathbb{P}_\Delta$ corresponds to the function h equal to 1 on the boundary of Δ^* and piecewise linear. The vertices of the triangulation of Δ^* defining Σ are by definition precisely the points of $N \cap \partial \Delta^*$ and h equals 1 at each of these points. These points generate the cones of dimension 1 in Σ over $\mathbb{Q}$. Since h is locally defined by an equation with integral coefficients, these points are actually the integral generators of cones of dimension 1 in Σ. It follows that h is indeed the support function of Σ and that D_h is the canonical divisor of $\mathbb{P}_\Sigma$. □

The usefulness of the preceding resides in the following lemma.

LEMMA 4.16. *The variety P_Σ is smooth in codimension 3.*

We use Lemma 4.2. Since the open sets U_σ defined by cones σ of dimension less than or equal to 3 cover the complement of a subset of codimension 4, it suffices to verify that these open sets are smooth, which is a consequence of the following lemma.

LEMMA 4.17. *Let e_1, e_2, e_3 be points of $\mathbb{Z}^n$ independent over $\mathbb{Q}$; assume that there exists a linear integral function h on $\mathbb{Z}^n$ such that $h(e_i) = 1$ and the only points of $\mathbb{Z}^n$ contained in the convex hull of the e_i are the points e_i. It is then possible to complete e_1, e_2, e_3 to a basis of $\mathbb{Z}^n$.*

Lemma 4.17 implies Lemma 4.16 by Lemma 4.2. For, by definition of Σ, all the integral points of $\partial \Delta^*$ are vertices of the triangulation, and all the cones of dimension 3 in Σ are cones on the convex hull of the three points $e_i \in N \cap \partial \Delta^*$ satisfying necessarily the hypotheses of Lemma 4.17.

The proof of Lemma 4.17 is elementary. It suffices to see that if we have a relation $\sum_i \alpha_i e_i = mu$ with $\alpha_i, m \in \mathbb{Z}^n$ and $u \in \mathbb{Z}^n$, then m divides α_i. The assumptions that $h(e_i) = 1$ and that h takes integral values yield $\sum_i \alpha_i = mh(u)$ with $h(u) \in \mathbb{Z}$. We note that we could replace α_i by $\alpha_i + m\beta_i$ with $\beta_i \in \mathbb{Z}$, so that α_i/m is defined modulo $\mathbb{Z}$ and satisfies $\sum_i (\alpha_i/m) e_i \in \mathbb{Z}^n$. Knowing that $\sum_i \alpha_i/m$ is an integer, we see immediately that by adding integers to the α_i/m or by changing the sign of all the α_i's, one may impose the conditions $\alpha_i/m \geq 0$, and $\sum_i \alpha_i/m = 1$, and at least one of the α_i/m is not integral if m does not divide the α_i, which produces an integral point different from the vertices in the convex hull of the e_i. $\square$

REMARK 4.18. In fact, even in higher dimension, the partial desingularization constructed above has some very good properties. It is a "quasi-smooth" variety (see [**63**]), which means that it is locally representable as the quotient of a smooth variety over a finite Abelian group, and its generic hypersurfaces, though singular, have pure Hodge structures on their cohomology groups (see [**68**]).

In what follows, we shall consider generic hypersurfaces with trivial canonical bundle $Z_f \subset \mathbb{P}_\Delta$ for f in $H^0(-K_{\mathbb{P}_\Delta})$ and their partial desingularization $\widehat{Z_f} = \tau^{-1}(Z_f)$, which is a hypersurface of $\mathbb{P}_\Sigma$. By the adjunction formula and Lemma 4.15, $\widehat{Z_f}$ has trivial canonical bundle. We shall show how to calculate the Hodge numbers $h^{1,1}(\widehat{Z_f})$ and $h^{n-2,1}(\widehat{Z_f})$, which makes sense by Remark 4.18. If one desires to work only with smooth varieties, one may assume $n = 4$ in what follows.

6. Calculation of the cohomology of $\widehat{Z_f}$

6.1. The results of Danilov and Khovanskii. Let $\mathbb{P}_\Delta$ be a polarized toric variety associated with a convex polyhedron having integral vertices $\Delta \subset M_\mathbb{R}$ (see 3.3). Let $Z_0 \subset U_0 \cong (\mathbb{C}^*)^n$ be a hypersurface obtained by restricting a divisor Z of $\mathbb{P}_\Delta$ with $Z \in |\mathcal{O}_{\mathbb{P}_\Delta}(1)|$.

DEFINITION 4.19. We say that Z_0 is Δ-*regular* if Z has a transversal intersection with all the strata of $\mathbb{P}_\Delta$ (see 1.2), that is, it has a smooth intersection with all the orbits O_σ of $\mathbb{P}_\Delta$.

For every algebraic variety X the cohomology groups of compact support $H^k_c(X)$ are endowed with mixed Hodge structures (see [**65**]). We thus have the notion of Hodge number $h^{p,q}(H^k_c(X)) = h^{p,q}(\mathrm{Gr}_W^{p+q}(H^k_c(X)))$, and we set

$$(4.12) \qquad e_{p,q}(X) = \sum_k (-1)^k h^{p,q}\big(H^k_c(X)\big).$$

The result of Danilov and Khovanskii [**64**] consists of the computation of $e_{p,q}(Z_0)$ as a function of the polyhedron Δ. This computation provides all the Hodge numbers of the smooth regular hypersurfaces of toric varieties, because of the following properties of $e_{p,q}$.

PROPOSITION 4.20. (i): *If X is smooth and compact, we have the equality* $e_{p,q}(X) = (-1)^{p+q} h^{p,q}(X)$.

(ii): If X is stratified by varieties X_i that are locally closed, we have
$$e_{p,q}(X) = \sum_i e_{p,q}(X_i).$$

Property (i) is immediate, since if X is smooth and compact, its mixed Hodge structures are pure, and we then have $h^{p,q}(H_c^k(X)) = 0$ for $k \neq p+q$.

Property (ii) is proved by induction on the number of strata and by using the long exact cohomology sequence with compact support. $\square$

Danilov and Khovanskii first calculate the number $\sum_q e_{p,q}(Z_0)$. To do this, they work with a refinement of Σ that provides a resolution of the singularities $\tau : \mathbb{P}_{\Delta'} \to \mathbb{P}_\Delta$, such that the following properties hold:
  (i): the divisor $D := \mathbb{P}_{\Delta'} - U_0$ is a divisor with normal crossings;
  (ii): the variety $Z' = \tau^{-1}(Z)$ is smooth, and $D_{Z'} := D \cap Z'$ is a divisor with normal crossings in Z'.

The existence of such a resolution is shown in [**63**].

The second property follows from the fact that Z_0 is Δ-regular. The calculation of $\sum_q e_{p,q}(Z_0)$ is then carried out as follows: we consider the complex $\Omega^\bullet_{Z',D_{Z'}}$ given by

(4.13) $$\Omega^\bullet_{Z',D_{Z'}} = \mathrm{Ker}\left(\Omega^\bullet_{Z'} \longrightarrow \Omega^\bullet_{D_{Z'}}\right).$$

This makes it possible to calculate the cohomology of compact support of $Z_0 = Z' - D_{Z'}$ and its Hodge filtration from the fact that $D_{Z'}$ has normal crossings (see [**65**]):

(4.14) $$\begin{cases} H_c^k(Z_0) = \mathbb{H}^k(\Omega^\bullet_{Z',D_{Z'}}), \\ F^p H_c^k(Z_0) = \mathbb{H}^k\big(0 \to \Omega^p_{Z',D_{Z'}} \to \cdots \to \Omega^{n-1}_{Z',D_{Z'}} \to 0\big). \end{cases}$$

From this we deduce immediately:
$$\sum_q e_{p,q}(Z_0) = (-1)^p \chi(Z', \Omega^p_{Z',D_{Z'}}).$$

We have in addition a series of exact sequences:

(4.15) $$0 \to \Omega^{p-1}_{Z',D_{Z'}}(-Z') \to \Omega^p_{\mathbb{P}_{\Delta'}, D_{|Z'}} \to \Omega^p_{Z',D_{Z'}} \to 0$$

which yield a resolution of $\Omega^p_{Z',D_{Z'}}$ by the sheaves $\Omega^{p+k}_{\mathbb{P}_{\Delta'},D_{|Z'}}(kZ')$, and hence the formula

(4.16) $$\chi(Z', \Omega^p_{Z',D_{Z'}}) = \sum_{k \geq 0} (-1)^k \chi\big(\Omega^{p+k+1}_{\mathbb{P}_{\Delta'},D_{|Z'}}((k+1)Z')\big)$$
$$= \sum_{k \geq 0} (-1)^k \big(\chi(\Omega^{p+k+1}_{\mathbb{P}_{\Delta'},D}((k+1)Z')) - \chi(\Omega^{p+k+1}_{\mathbb{P}_{\Delta'},D}(kZ'))\big).$$

We denote by $l^*(\Delta)$ the number of integral points inside Δ. The following proposition expresses the Euler–Poincaré characteristics $\chi\big(\Omega^{p+k+1}_{\mathbb{P}_{\Delta'},D}(kZ')\big)$ as a function of $l^*(k\Delta)$.

PROPOSITION 4.21 (see [**64**]). *The following relations hold:*
$$\chi\big(\mathbb{P}_{\Delta'}, \Omega^p_{\mathbb{P}_{\Delta'},D}(kZ')\big) = C_p^n l^*(k\Delta) \quad \text{for} \quad k > 0,$$
$$\chi\big(\mathbb{P}_{\Delta'}, \Omega^p_{\mathbb{P}_{\Delta'},D}\big) = (-1)^n C_p^n.$$

6. CALCULATION OF THE COHOMOLOGY OF $\widehat{Z}_f$

Formula (4.16) thus becomes

$$(4.17) \quad \chi(Z', \Omega^p_{Z', D_{z'}}) = (-1)^p \sum_q e_{p,q}(Z_0)$$

$$= (-1)^{n+1} C^n_{p+1} + \sum_{k \geq 1} (-1)^k C^n_{p+1+k} \big(l^*((k+1)\Delta) - l^*(k\Delta)\big)$$

$$+ C^m_{p+1} l^*(\Delta).$$

which can be immediately rewritten as

$$(4.18) \quad (-1)^p \sum_q e_{p,q}(Z_0) = \sum_{k \geq 0} (-1)^k C^{n+1}_{p+2+k} l^*((k+1)\Delta) + (-1)^{n+1} C^n_{p+1}.$$

To conclude, we use Theorem 4.22 stated below, of Lefschetz type for toric varieties, and induction on the dimension to show that knowledge of $\sum_q e_{p,q}(Z_0)$ suffices to determine $e_{p,q}(Z_0)$.

THEOREM 4.22 (see [**64**]). *Let $Z_0 \subset (\mathbb{C}^*)^n$ be a Δ-regular hypersurface. Then*

$$(4.19) \quad \begin{cases} h^{p,q}\big(H^k_c(Z_0)\big) = 0 & \text{for } k < n - 1, \\ h^{p,q}\big(H^{n-1}_c(Z_0)\big) = 0 & \text{for } p + q > n - 1, \\ h^{p,q}\big(H^k_c(Z_0)\big) = h^{p,q}\big(H^{k+2}_c((\mathbb{C}^*)^n)\big) & \text{for } k > n - 1. \end{cases}$$

The first equality follows from the fact that Z_0 is an affine variety of dimension $(n-1)$ (see [**67**]). The second follows from the fact that Z_0 is smooth. □

The theorem yields $e_{p,q}(Z_0)$ for $p + q > n - 1$. We now take a smooth compactification Z' of Z_0 as above. We have a stratification of Z' by toric affine varieties Z'_σ (see 1.2), since the open stratum equals Z_0. By induction on the dimension we may assume that the $e_{p,q}(Z'_\sigma)$ are known for all other strata. On the other hand, by Proposition 4.20 (*ii*) we have

$$(4.20) \quad e_{p,q}(Z') = e_{p,q}(Z_0) + \sum_{\sigma \neq \{0\}} e_{p,q}(Z'_\sigma).$$

We therefore know $e_{p,q}(Z')$ for $p + q > n - 1$. But Z' is smooth and compact, and we thus also know $e_{p,q}(Z')$ for $p + q < n - 1$ by Poincaré duality. Thus we know $e_{p,q}(Z_0)$ for $p + q < n - 1$ (by formula (4.20) and assuming known the $e_{p,q}(Z'_\sigma)$, $\sigma \neq \{0\}$). We can therefore deduce the value of $e_{p,q}(Z_0)$ from $\sum_q e_{p,q}(Z_0)$.

The calculation just described does not in general yield an explicit algorithm for obtaining $e_{p,q}(Z_0)$. One can, however, deduce from it the value of $e_{n-2,1}(Z_0)$ as follows: the formula (4.18) for $p = n - 2$, combined with Theorem 4.22, yields the equality

$$(4.21) \quad (-1)^{n-2}\big(e_{n-2,1}(Z_0) + e_{n-2,0}(Z_0)\big) + (-1)^{n-1} C^n_{n-1}$$

$$= (-1)^{n+1} C^n_{n-1} - l^*(2\Delta) + (n+1)l^*(\Delta),$$

where we have used the equality (Theorem 4.22)

$$e_{n-2,n-2}(Z_0) = e_{n-1,n-1}(\mathbb{C}^*)^n = -C^n_{n-1},$$

that is,

$$(4.22) \quad e_{n-2,1}(Z_0) + e_{n-2,0}(Z_0) = (-1)^{n-1} l^*(2\Delta) + (-1)^n (n+1) l^*(\Delta).$$

Danilov and Khovanskii then prove the following proposition using the reasoning described above.

PROPOSITION 4.23. *The following relations hold*
$$h^{n-1,0}\big(H_c^{n-1}(Z_0)\big) = (-1)^{n-1}e_{n-1,0}(Z_0) = l^*(\Delta),$$
$$h^{n-2,0}\big(H_c^{n-1}(Z_0)\big) = (-1)^{n-1}e_{n-2,0}(Z_0) = \sum_{\text{codim}\,\theta=1} l^*(\theta).$$

In the second formula θ ranges over faces of codimension 1 of the initial polyhedron Δ. One can thus finally express the Hodge number $h^{n-2,1}\big(H_c^{n-1}(Z_0)\big)$ as a function of the numerical characteristics of the polyhedron Δ for $n \geq 4$ since by Theorem 4.22
$$h^{n-2,1}\big(H_c^{n-1}(Z_0)\big) = (-1)^{n-1}e_{n-2,1}(Z_0) \quad \text{for} \quad n \geq 4.$$

COROLLARY 4.24. *The following relation holds:*
$$(4.23) \qquad h^{n-2,1}(Z_0) = l^*(2\Delta) - (n+1)l^*(\Delta) - \sum_{\text{codim}\,\theta=1} l^*(\theta).$$

REMARK 4.25. When Δ is a reflexive polyhedron, its faces of codimension 1 are described by equations $\langle m^*, \cdot \rangle = 1$, where $m^* \in N$ is an integer, since the dual polyhedron has integer vertices. The integral points u interior to 2Δ satisfy the inequality $\langle m^*, u \rangle < 2$ for all the faces of codimension 1 of Δ, and hence $\langle m^*, u \rangle \leq 1$. From this we conclude that $l^*(2\Delta) = l(\Delta)$, where $l(\Delta)$ is the number of integral points contained in Δ and $l^*(\Delta) = 1$ for similar reasons.

6.2. Batyrev's calculation. We now return to the hypersurfaces $\widehat{Z}_f \subset \mathbb{P}_\Sigma$ of dimension $(n-1)$ having trivial canonical bundle, and we calculate $h^{n-2,1}(\widehat{Z}_f) = (-1)^{n-1}e_{n-2,1}(\widehat{Z}_f)$. By using Proposition 4.20 we see that
$$(4.24) \qquad e_{n-2,1}(\widehat{Z}_f) = \sum_{\sigma'} e_{n-2,1}(\widehat{Z}_{f_{\sigma'}}),$$
where σ' ranges over all the cones of the refinement Σ (see 5), and $\widehat{Z}_{f_{\sigma'}}$ is the intersection of $\widehat{Z}_f$ with the orbit corresponding to σ' (see 1.2).

It is thus a matter of describing the components of the boundary $\widehat{Z}_f - \widehat{Z}_{f_0}$.

We recall that the fan Σ is obtained by triangulation of Δ^* and satisfies the condition that the generators of cones of dimension 1 in Σ are the integer points of $\Delta^* - \{0\}$. For each cone of dimension 1 in Σ with generator $v \in N$, there thus exists a unique face θ^* of Δ^* such that v belongs to the interior of θ^*. On the other hand, by 1.2, to v there corresponds a divisor D_v of $\mathbb{P}_\Sigma$ having a dense open set isomorphic to $(\mathbb{C}^*)^{n-1}$ (which is the orbit $O_{\mathbb{R}_v^+}$, denoted from now on by O_v) and to this face θ^* there corresponds the cone $\widetilde{\theta}^*$ over θ^* that belongs to the system of cones that define $\mathbb{P}_\Delta$, hence a stratum $O_{\widetilde{\theta}^*}$ of $\mathbb{P}_\Delta$ isomorphic to $(\mathbb{C}^*)^{n-k}$, where $\dim \theta^* = k - 1$. We therefore have the following lemma.

LEMMA 4.26. *The map $\tau : \mathbb{P}_\Sigma \to \mathbb{P}_\Delta$ satisfies $\tau(O_v) = O_{\widetilde{\theta}^*}$.*

Indeed, this follows from the definitions of the strata O_v and $O_{\widetilde{\theta}^*}$. Since v is in θ^*, by 5, the map τ sends $\text{Spec}\, A_v$ to $\text{Spec}\, A_{\widetilde{\theta}^*}$. But
$$A_v = \mathbb{C}[X^m]_{\langle m,v \rangle \geq 0}$$

and O_v is defined by the X^m such that $\langle m, v \rangle = 1$. On the other hand, $O_{\widetilde{\theta^*}} \subset$ Spec $A_{\widetilde{\theta^*}}$ is defined by the equations X^m for $m \in \widetilde{\theta}^{*^*} - \widetilde{\theta}^{*^\perp}$. Since v is an interior point of θ^*, we have $m(v) > 0$ for all $m \in \widetilde{\theta}^{*^*} - \widetilde{\theta}^{*^\perp}$, and thus X^m vanishes on O_v. We thus have $\tau(O_v) \subset O_{\widetilde{\theta^*}}$, and the equality follows from the fact that T acts transitively and compatibly with τ on each of these varieties. □

Lemma 4.26 shows immediately that $O_v \cap \widehat{Z_f}$ is fibered in $(\mathbb{C}^*)^k$ over $Z_{f,\widetilde{\theta^*}}$ for the face θ^* of Δ^* of dimension k such that v is interior to θ^*. It is in addition clear by the T-equivariance of τ that all the strata of $\widehat{Z_f}$ are fibered in $(\mathbb{C}^*)^l$ over the strata of Z_f. Since the cohomology with compact supports of $(\mathbb{C}^*)^l$ satisfies

$$h^{p,q}\big(H_c^k(\mathbb{C}^{*l})\big) = 0 \quad \text{for} \quad p \neq q,$$

we see that a stratum of $\widehat{Z_f}$ can contribute to $e_{n-2,1}(\widehat{Z_f})$ only if the corresponding stratum of Z_f satisfies $e_{n-3,0} \neq 0$ or $e_{n-2,1} \neq 0$. But, by Theorem 4.22, the last inequality is possible only for the open stratum, whose contribution is already known. Similarly, the first inequality is possible only for a stratum $Z_{f,\widetilde{\theta^*}}$ of dimension at least $(n-3)$.

Thus for a stratum $\widehat{Z_f}_\sigma$ over $Z_{f,\widetilde{\theta^*}}$ to contribute to $e_{n-2,1}(\widehat{Z_f})$ it is necessary that its fibers over $Z_{f,\widetilde{\theta^*}}$ be of dimension at least 1. We have thus finally found that the only strata of $\widehat{Z_f}$ that contribute to $e_{n-2,1}(\widehat{Z_f})$ are the components of O_v, where v is an interior point of θ^* with $\dim \theta^* = 1$. Since we have

$$h^{0,0}\big(H_c^1(\mathbb{C}^*)\big) = 1 = h^{1,1}\big(H_c^2(\mathbb{C}^*)\big)$$

and $\tau^{-1}(Z_{f,\widetilde{\theta^*}})^0$ has exactly $l^*(\theta^*)$ components fibered in $\mathbb{C}^*$ over $Z_{f,\widetilde{\theta^*}}$ for $\dim \theta^* = 1$, Lemma 4.26 shows that the contribution of the dense open set fibered in $\mathbb{C}^*$

$$\tau^{-1}(Z_{f,\widetilde{\theta^*}})^0 \subset \tau^{-1}(Z_{f,\widetilde{\theta^*}})$$

to $e_{n-2,1}(\widehat{Z_f})$ is

$$l^*(\theta^*) e_{n-3,0}(Z_{f,\widetilde{\theta^*}}) = (-1)^{n-1} l^*(\theta^*) h^{n-3,0}(Z_{f,\widetilde{\theta^*}}).$$

Proposition 4.23 then finally yields

(4.25) $$h^{n-3,0}(Z_{f,\widetilde{\theta^*}}) = l^*(\theta),$$

where $\theta \subset \Delta$ is the dual face of θ^*. (It should be noted here that the closure of $O_{\widetilde{\theta^*}}$ in $\mathbb{P}_\Delta$ is isomorphic to $\mathbb{P}_\theta$.) We have thus shown:

PROPOSITION 4.27. *The Hodge number $h^{n-2,1}(\widehat{Z_f})$ is calculated by the formula*

(4.26) $$h^{n-2,1}(\widehat{Z_f}) = l(\Delta) - (n+1) - \sum_{\text{cod } \theta = 1} l^*(\theta) + \sum_{\text{cod } \theta = 2} l^*(\theta^*) l^*(\theta),$$

where we have identified the faces θ of Δ having dimension 2 with the faces θ^* of Δ^* having codimension 1.

6.3. The calculation of $h^{1,1}(\widehat{Z_f})$. Continuing to follow [60] and [61], we explain how to calculate $h^{1,1}(\widehat{Z_f})$. Assume $n \geq 4$; we then have $h^{2,0}(\widehat{Z_f}) = 0$ and hence rank $\text{Pic}_\mathbb{Q}(\widehat{Z_f}) = h^{1,1}(\widehat{Z_f})$. Batyrev shows the following result.

PROPOSITION 4.28. *The $\mathbb{Q}$-vector space $\text{Pic}_\mathbb{Q}(\widehat{Z_f})$ is generated by the classes of components of $\widehat{Z_f} - Z_{f,0}$, where $Z_{f,0}$ is the affine part $\widehat{Z_f} \cap (\mathbb{C}^*)^n$ of $\widehat{Z_f}$.*

On the other hand, these components are not independent. Indeed, the group that they generate contains the restrictions of the T-invariant divisors of $\mathbb{P}_\Sigma$. (We note that there is no longer space here to distinguish the Weil divisors from the Cartier divisors of $\mathbb{P}_\Sigma$ over $\mathbb{Q}$, since all the cones of Σ are simplicial by construction, that is, generated by independent vertices, and hence the linear functions with values in $\mathbb{Q}$ over these cones are exactly determined by their values on the vertices.)

But, we have noted in Remark 4.4 that there exist n relations on $\operatorname{Pic}_\mathbb{Q}(\mathbb{P}_\Sigma)$ given by the fact that the globally linear entire functions over N (defined by an element of M) correspond to principal Cartier divisors. Batyrev shows that these relations generate all the relations between the components of the boundary. We thus find

$$h^{1,1}(\widehat{Z_f}) = \operatorname{rank} \operatorname{Pic}_\mathbb{Q}(\widehat{Z_f}) = (\text{number of components of } \widehat{Z_f} - Z_f^0) - n.$$

We now use Lemma 4.26, which shows that the divisors D_v of $\mathbb{P}_\Sigma$ have a dense open set O_v that is fibered in $(\mathbb{C}^*)^k$ over an orbit $O_{\widetilde{\theta^*}}$ of $\mathbb{P}_\Delta$ for θ^* a face of Δ^* having dimension k.

- If $k+1 = n$, the orbit $O_{\widetilde{\theta^*}}$ is of dimension 0 and hence does not intersect Z_f. Each of the strata $O_{\widetilde{\theta^*}}$ provides an additional $l^*(\theta^*)$ components of $\mathbb{P}_\Sigma - U_0$, since the interior points of θ^* provide precisely (by Lemma 4.26) the T-invariant divisors of $\mathbb{P}_\Sigma$ contracted over $O_{\widetilde{\theta^*}}$.
- If $k+1 = n-1$, the orbit $O_{\widetilde{\theta^*}}$ is of dimension 1 and meets Z_f in exactly $l^*(\theta) + 1$ points, where $\theta \subset \Delta$ is the dual face of θ^*, as follows from Lemma 4.7.
- Finally, if $k+1 < n-1$, the intersection $O_{\widetilde{\theta^*}} \cap Z_f$ is irreducible and non-empty by amplitude.

From this one can deduce that the number of components of $\widehat{Z_f} - Z_f^0$ is given by the sum

$$\begin{cases} l(\Delta^*) - 1 & (\text{number of components of } \mathbb{P}_\Sigma - U_0) \\ - \sum_{\dim \theta^* = n-1} l^*(\theta^*) & (\text{corresponding to divisors } D_v \text{ that do not meet } \widehat{Z_f}) \\ + \sum_{\dim \theta^* = n-2} l^*(\theta) l^*(\theta^*) & (\text{corresponding to divisors } D_v \text{ whose intersection} \\ & \text{with } \widehat{Z_f} \text{ is reducible}). \end{cases}$$

We have thus shown the following:

PROPOSITION 4.29. *The number $h^{1,1}(\widehat{Z_f})$ is calculated from the formula*

$$(4.27) \quad h^{1,1}(\widehat{Z_f}) = l(\Delta^*) - 1 - n - \sum_{\dim \theta^* = n-1} l^*(\theta^*) + \sum_{\dim \theta^* = n-2} l^*(\theta) l^*(\theta^*)$$

where, as above, θ^* is a face of Δ^* and θ is the dual face of Δ.

Comparing propositions 4.28 and 4.29, we see immediately that $h^{1,1}(\widehat{Z_f})$ is obtained by replacing Δ with Δ^* in the expression for $h^{n-2,1}(\widehat{Z_f})$. We have thus shown the following theorem.

THEOREM 4.30. *The involution $\Delta \mapsto \Delta^*$ between reflexive polyhedra interchanges the Hodge numbers $h^{1,1}(\widehat{Z_f})$ and $h^{n-2,1}(\widehat{Z_{f^*}})$, where $Z_f \subset \mathbb{P}_\Delta$ is defined by a generic section of $-K_{\mathbb{P}_\Delta}$ and $Z_{f^*} \subset \mathbb{P}_{\Delta^*}$ is defined by a generic section of $-K_{\mathbb{P}_{\Delta^*}}$, in accordance with the mirror property (1.47).*

REMARK 4.31. We have nevertheless not constructed mirror symmetry as an involution between the families $\{\widehat{Z_f}\}$ and $\{\widehat{Z_{f^*}}\}$, since the deformation class of $\widehat{Z_f}$ depends on the choice of the desingularization $\tau : \mathbb{P}_\Sigma \to \mathbb{P}_\Delta$.

CHAPTER 5

Quantum cohomology

The present chapter begins by describing the axiomatic formulation of the Gromov–Witten invariants due to Kontsevich and Manin, which is of interest because it reveals the "operadic" structure, defining them as a series of polynomial invariants with values in the cohomology of $\overline{M}_{g,n}$ and exhibiting the naturality of the properties that they satisfy, which reflect the existence of certain universal operations on the $\overline{M}_{g,n}$ (restriction to the boundary, forgetting a point, etc.).

Still following [78], we explain how the WDVV equation, which is satisfied by the Gromov–Witten potential, can be derived from these axioms. We also describe, following Dubrovin, the connection between this equation (satisfied by a function on a manifold M endowed with coordinates) and the flatness of a certain connection on the tangent bundle to M constructed using this function. We then apply these considerations to the construction of a complex variation of Hodge structure parameterized by an open set of $H^2(X, \mathbb{C})$, where X is a Calabi–Yau threefold with $h^{2,0} = 0$.

We explain finally some results that are theoretically less ambitious, yet sufficient for the applications mentioned above, obtained by Gromov, Ruan, and Ruan-Tian: the definition (depending only on the assignment of a symplectic structure) of a restricted version of the Gromov–Witten invariants in terms of counting of rational curves, and the proof of the crucial splitting property.

The chapter ends with a section devoted to the formula of Aspinwall and Morrison, which computes the contribution to quantum cohomology of the (large-dimension) family of branched coverings of degree k of a rigid rational curve on a Calabi–Yau threefold. This formula makes it possible to calculate the quantum cohomology of a Calabi–Yau threefold all of whose generically embedded rational curves are rigid, as a function of the number of the latter in each homology class.

1. The formulation by Kontsevich and Manin

Kontsevich and Manin [78] have given the most general definition of Gromov–Witten invariants in an axiomatic manner, and in this generality their existence and the properties they are supposed to satisfy have not been rigorously established.

Given an algebraic variety V (or more generally a symplectic manifold), these invariants should be described by a series of maps

(5.1) $$H_{A,g,k} : H^*(V^k) \to H^*(\overline{M}_{g,k})$$

for $A \in H_2(V, \mathbb{Z})$ and g and k integers such that $g \geq 2$, or $g = 1$ and $k \geq 1$, or $g = 0$ and $k \geq 3$, so that the moduli space $\overline{M}_{g,k}$ of curves of genus t with k stable marked points (see [74]) is well defined. We use the notation

$$c_1(V) = c_1(-K_V) \in H^2(V, \mathbb{Z}),$$

which is defined in all the cases considered above (see 1.3) since the simple data of a symplectic structure determine a class of deformations of pseudocomplex structures (see [**76**], [**86**]).

1.1. Virtual construction. Suppose that for every A, g, and k, as above, we have a moduli space $\mathrm{Mor}_A(g, V, k)$ of k-pointed holomorphic curves of genus g and class A in V, with the "correct" real dimension (that is, computed by the Riemann–Roch formula)
$$d = 2\big(c_1(V)A - (g-1)(n-3) + k\big), \quad n = \dim_{\mathbb{C}} V.$$
We think of $\mathrm{Mor}_A(g, V, k)$ as the set of assignments $(\phi, x_1, \ldots, x_k)$ where $\phi : C \to V$ is a holomorphic map such that $\phi_*([C]) = A$, C is a curve of genus g, and the x_i are points of C, modulo the action of the automorphisms ψ of C:
$$\psi\big((\phi, x_1, \ldots, x_k)\big) = \big(\phi \circ \psi, \psi^{-1}(x_1), \ldots, \psi^{-1}(x_k)\big).$$
Thus on the one hand we have a classifying map
$$\pi : \mathrm{Mor}_A(g, V, k) \to \overline{\mathcal{M}}_{g,k}$$
(which one would like to be proper), and on the other hand an evaluation map

(5.2) $$\begin{cases} \mathrm{ev} : \mathrm{Mor}_A(g, V, k) \to V^k, \\ (\phi, x_1, \ldots, x_k) \mapsto \big(\phi(x_1), \ldots, \phi(x_k)\big) \end{cases}$$

and the map $H_{A,g,k}$ should then be defined by:

(5.3) $$H_{A,g,k}(\alpha) = \pi_* \circ \mathrm{ev}^*(\alpha).$$

The construction of $\mathrm{Mor}_A(g, V, k)$ is problematic, especially since certain components of the Hilbert scheme of the curves of genus g and class A on V do not have the correct dimension (see 6).

In any case the Hilbert scheme does not really solve the problem, since it does not parameterize a family of stable curves.

We shall explain in § 2 how this construction can be partially realized by studying the solutions of the "Cauchy–Riemann equation with an inhomogeneous term" for a generic pseudocomplex structure on V (see [**81**], [**82**], [**83**]). The fundamental result of [**83**] makes it possible essentially to calculate the expressions $\int_{\overline{\mathcal{M}}_{g,k}} H_{A,g,k}(\alpha)$.

1.2. Axioms. (A1) For $\alpha \in H^l(V^k)$ we have:
$$\deg\big(H_{A,g,k}(\alpha)\big) = l + 2\big(-(c_1(V) \cdot A) + (g-1)n\big).$$

(A2) The map $H_{A,g,k}$ is equivariant for the natural action of the symmetric group S_k on V^k and $\overline{\mathcal{M}}_{g,k}$.

(A3) Let 1_V be the canonical generator of $H^0(V)$; then if (g, k) is such that $(g, k-1)$ is in the set considered above, so that the map that forgets the kth point $\pi_k : \overline{\mathcal{M}}_{g,k} \to \overline{\mathcal{M}}_{g,k-1}$ is defined, we have:

(5.4) $$H_{A,g,k}(\alpha_1 \otimes \cdots \otimes \alpha_{k-1} \otimes 1_V) = \pi_k^*\big(H_{A,g,k-1}(\alpha_1 \otimes \cdots \otimes \alpha_{k-1})\big).$$

In addition $H_{A,0,3}(\alpha_1 \otimes \alpha_2 \otimes 1_V)$ should equal $\int_V \alpha_1 \wedge \alpha_2$ for $A = 0$ and 0 otherwise.

(A4) Let $\beta \in H^2(V)$ be the class of a divisor; then when π_k is defined, we have

(5.5) $$\pi_{k*}\big(H_{A,g,k}(\alpha \otimes \cdots \otimes \alpha \otimes \beta)\big) = (\beta \cdot A) H_{A,g,k-1}(\alpha \otimes \cdots \otimes \alpha)$$

(A5) When $g = 0$ and $A = 0$ (so that we are considering constant (pseudo)holomorphic maps of $\mathbb{P}^1$ into V), we have

(5.6) $\quad H_{0,0,k}(\alpha_1 \otimes \cdots \otimes \alpha_k) = \begin{cases} 0, & \text{if } \sum_i \deg \alpha_i \neq 2n, \\ \left(\int_V \alpha_1 \wedge \cdots \wedge \alpha_k \right) 1_{\overline{\mathcal{M}}_{0,k}} & \text{otherwise.} \end{cases}$

(A6) Let $g_1 + g_2 = g$ and $k_1 + k_2 = k$. Then we have a map

$$\phi : \overline{\mathcal{M}}_{g_1,k_1+1} \times \overline{\mathcal{M}}_{g_2,k_2+1} \to \overline{\mathcal{M}}_{g,k}$$

which associates with $(C_1, x_1, \ldots, x_{k_1+1})$ and $(C_2, y_1, \ldots, y_{k_2+1})$ the curve $C = C_1 \bigcup_{x_{k_1+1}=y_{k_2+1}} C_2$ with the marked points $x_1, \ldots, x_{k_1}, y_1, \ldots, y_{k_2}$. Thus we then ask

(5.7) $\quad \phi^* \big(H_{A,g,k}(\alpha_1 \otimes \cdots \otimes \alpha_k) \big)$

$$= \sum_{\substack{A_1+A_2=A \\ \sigma, \tau}} g^{\sigma\tau} H_{A_1,g_1,k_1+1}(\alpha_1 \otimes \cdots \otimes \alpha_{k_1} \otimes e_\sigma)$$

$$\otimes H_{A_2,g_2,k_2+1}(\alpha_{k_1+1} \otimes \cdots \otimes \alpha_k \otimes e_\tau)$$

where e_σ is a basis of $H^*(V)$ and $g^{\sigma\tau}$ is the matrix inverse to the intersection matrix $\langle e_\sigma, e_\tau \rangle$.

1.3. Heuristic justification of the axioms. • The axiom A1 is an immediate consequence of the calculation of the virtual dimension of $\text{Mor}_A(g, V, k)$ (see 1.1) and the dimension of $\overline{\mathcal{M}}_{g,k}$, which is $3g - 3 + k$.

• The axiom A2 results from equivariance relative to S_k of the evaluation diagram

(5.8) $\quad \begin{array}{ccc} \text{Mor}_A(g, V, k) & \xrightarrow{\text{ev}} & V^k \\ {\scriptstyle \pi} \downarrow & & \\ \overline{\mathcal{M}}_{g,k} & & \end{array}$

• The axiom A3 comes from the fact that if π_k is defined, we have three arrows that forget the kth point

$$\pi_k : V^k \to V^{k-1}, \quad \pi_k : \text{Mor}_A(g, V, k) \to \text{Mor}_A(g, V, k-1), \quad \pi_k : \overline{\mathcal{M}}_{g,k} \to \overline{\mathcal{M}}_{g,k-1}$$

which makes the preceding diagram the "pull-back" of the evaluation diagram

(5.9) $\quad \begin{array}{ccc} \text{Mor}_A(g, V, k-1) & \xrightarrow{\text{ev}} & V^{k-1} \\ {\scriptstyle \pi} \downarrow & & \\ \overline{\mathcal{M}}_{g,k-1} & & \end{array}$

But a class $\alpha \in H^*(V^k)$ is of the form $\beta \otimes 1_V$ with $\beta \in H^*(V^{k-1})$ if and only if it is equal to $\pi_k^* \beta$. The preceding implies immediately that $H_{A,g,k}(\alpha) = \pi_k^* \big(H_{A,g,k-1}(\beta) \big)$. In addition, in the case when $g = 0$ and $k = 3$, $\overline{\mathcal{M}}_{0,3}$ is reduced to a point and we have

$$H_{A,g,k}(\alpha) = \int_{\text{Mor}_A(0,V,3)} \text{ev}^*(\alpha) \quad \text{for } \alpha \in H^*(V^3).$$

We note now that if $A \neq 0$, the map $\pi_2 \circ \text{ev} : \text{Mor}_A(0, V, 3) \to V^2$ has fibers of positive dimension, since the holomorphic maps $f : \mathbb{P}^1 \to V$ of class A are nonconstant. Thus for x_1, x_2, x_3 fixed in $\mathbb{P}^1$ and $x_3' \in \mathbb{P}^1$ distinct from x_3 we have $(f, x_1, x_2, x_3) \neq (f, x_1, x_2, x_3')$ in $\mathrm{Mor}_A(0, V, 3)$, which provides fibers of positive dimension. We thus clearly have

$$\int_{\mathrm{Mor}_A(0,V,3)} \mathrm{ev}^*(\pi_2^*\beta) = 0 \quad \text{for } \beta \in H^*(V^2).$$

In contrast, when $A = 0$, the map $\mathrm{ev} : \mathrm{Mor}_0(0, V, 3) \to V^3$ can be identified with the inclusion of the diagonal $\{(v, v, v), v \in V\} \subset V^3$ and hence the last assertion of A3 is clear.

- To see A4, we note that if H_{α_i} are cycles whose homology classes are Poincaré dual to α_i, then $H_{A,g,k}(\alpha_1 \otimes \cdots \otimes \alpha_k)$ is the cohomology class of the cycle

$$V_{A,H_{\alpha_1},\ldots,H_{\alpha_k}} = \{(C, x_1, \ldots, x_k); \exists f : C \to V, f_*([C]) = A \text{ and } f(x_i) \in H_{\alpha_i}\}.$$

Now assume that H_{α_k} is an effective divisor and that π_k is defined. It is then clear that π_k restricted to $V_{A,H_{\alpha_1},\ldots,H_{\alpha_k}}$ is finite, of degree $A \cdot \alpha_k$ on its image $V_{A,H_{\alpha_1},\ldots,H_{\alpha_{k-1}}}$ since for $(C, x_1, \ldots, x_{k-1})$ in $V_{A,H_{\alpha_1},\ldots H_{\alpha_{k-1}}}$ one can choose the point x_k so that $(C, x_1, \ldots, x_k)$ belongs to $V_{A,H_{\alpha_1},\ldots,H_{\alpha_k}}$ arbitrarily among the points of intersection of $f(C)$ and H_{α_k}.

- Finally Axiom A6 can be justified as follows. The notation is the same; we wish to calculate the intersection class of the cycle $V_{A,H_{\alpha_1},\ldots,H_{\alpha_k}}$ with

$$\phi(\overline{\mathcal{M}}_{g_1,k_1+1} \times \overline{\mathcal{M}}_{g_2,k_2+1}).$$

We see immediately that this is the union, over all decompositions of A into a sum $A_1 + A_2$, of the cycles formed from the couples

$$((C_1, x_1, \ldots, x_{k_1+1}), (C_2, y_1, \ldots, y_{k_2+1}))$$

such that there exist holomorphic maps $f_i : C_i \to V$ representing A_i for $i = 1, 2$ and such that $f_1(x_i) \in H_{\alpha_i}$, $f_2(y_j) \in H_{\alpha_{k_1+j}}$ and $f_1(x_{k_1+1}) = f_2(y_{k_2+1})$. The last condition can then be written

$$(f_1(x_{k_1+1}), f_2(y_{k_2+1})) \in \Delta \subset V \times V.$$

But the diagonal Δ of $V \times V$ is homologous to $\sum_{\sigma,\tau} g^{\sigma\tau} H_\sigma \times H_\tau$, where the H_σ are cycles of cohomology class e_σ. The set above is thus homologous to

$$\sum_{\sigma,\tau} g^{\sigma\tau} V_{A_1,H_{\alpha_1},\ldots,H_{\alpha_{k_1}},H_\sigma} \times V_{A_2,H_{\alpha_{k_1+1}},\ldots,H_{\alpha_k},H_\tau},$$

which "shows" Axiom A6.

2. The work of Ruan and Tian

2.1. Mixed Gromov–Witten invariants. We shall consider only the case of genus 0, although the works of Ruan and Tian involve pseudoholomorphic curves of arbitrary genus. Their results make it possible to construct the invariants

$$\int_{\overline{\mathcal{M}}_{0,n}} H_{A,0,n}(\alpha_1 \otimes \cdots \otimes \alpha_n)$$

(which are the only ones used in what follows), essentially for the symplectic varieties (V, ω) called *monotone*, that is, such that $c_1(V)$ is a positive or zero multiple of the class of ω.

More generally, for such manifolds, Ruan and Tian construct *mixed invariants* $\Phi_A(\alpha_1, \ldots, \alpha_k | \beta_1, \ldots, \beta_l)$ such that

$$\Phi_A(\alpha_1, \alpha_2, \alpha_3 | \beta_1, \ldots, \beta_l) = \int_{\overline{M}_{0,l+3}} H_{A,0,l+3}(\alpha_1 \otimes \alpha_2 \otimes \alpha_3 \otimes \beta_1 \otimes \cdots \otimes \beta_l)$$

and whose principal virtue, except for the (anti)symmetry properties with respect to the permutations of α or β (Proposition 2) is the following property:

(5.10) $\Phi_A(\alpha_1, \ldots, \alpha_k | \beta_1, \ldots, \beta_l)$

$$= \sum_{\substack{i_1 < \cdots < i_r, r \leq l \\ A_1 + A_2 = A}} \sum_{\sigma,\tau} g^{\sigma\tau} \Phi_{A_1}(\alpha_1, \ldots, \alpha_s, e_\sigma | \beta_{i_1}, \ldots, \beta_{i_r})$$

$$\times \Phi_{A_2}(\alpha_{s+1}, \ldots, \alpha_k, e_\tau | \beta_{j_1}, \ldots, \beta_{j_{l-r}}),$$

where e_σ is a basis of $H^{2*}(V)$, $g^{\sigma\tau}$ is the inverse of the intersection matrix, s is a fixed integer, and $j_1 < \cdots < j_{l-r}$ is the ordered set complementary to $\{i_1, \ldots, i_r\}$.

2.2. Description of the invariants. Let (V, ω) be a symplectic variety:
- we make the assumption
$$c_1(V) = \lambda[\omega] \in H^2(V, \mathbb{Z}), \quad \lambda \geq 0$$

- or the following weaker assumption, called the assumption of *weak monotony*: if $A \in H_2(V, \mathbb{Z})$ is such that $(\omega, A) > 0$ and $(c_1(V), A) \geq 3 - n$, then $(c_1(V), A) \geq 0$.

This assumption implies that if J is a pseudocomplex structure compatible with ω (which means that ω is J-invariant and $\omega(v, Jv) > 0$, $v \in TV$, $v \neq 0$), the nonconstant pseudoholomorphic rational curves $f : \mathbb{P}^1 \to V$ on V satisfy the condition

$$(c_1(V), f_*([\mathbb{P}^1])) \geq 0.$$

Indeed, for such a pseudoholomorphic curve, which one may assume generically embedded, we certainly have $(\omega, f_*([\mathbb{P}^1])) > 0$; and, since J is generic,

$$(c_1(V), f_*([\mathbb{P}^1])) \geq 3 - n,$$

because the dimension of the space of generically embedded pseudoholomorphic curves of class A given modulo the action of $\mathrm{Aut}\,\mathbb{P}^1$ is the expected one, that is, equal to

$$2((c_1(V), A) + n - 3)$$

(see [81], [86]) and this number must be positive or zero.

Tian and Ruan consider the maps $f : C \to V$ satisfying the Cauchy–Riemann equation with an inhomogeneous term. In the case when $g = 0$ and hence $C \cong \mathbb{P}^1$, this means that we fix on $\mathbb{P}^1 \times V$ a section ν of class $\mathcal{C}^\infty$ of the bundle $\mathcal{E}$ whose fiber at the point (t, v) is the space of $\mathbb{C}$-antilinear maps from $T_{\mathbb{P}^1,t}$ into $T_{V,v}$. For a differentiable map $f : \mathbb{P}^1 \to V$, one can construct $\bar\partial f$ (see 1.1), which is the $\mathbb{C}$-antilinear part of $df \in \mathrm{Hom}\,(T_{\mathbb{P}^1}, f^*T_V)$ and the Cauchy–Riemann equation with inhomogeneous term ν is then

(5.11) $$\bar\partial f = (\mathrm{Id}, f)^*\nu.$$

For fixed J and ν we denote by $W_{J,\nu,A}$ the set of maps f of $\mathbb{P}^1$ into V satisfying the equation (5.11) and such that $f_*([\mathbb{P}^1]) = A$.

It can be shown that for (J,ν) generic, $W_{J,\nu,A}$ is smooth, naturally oriented (see [**81**]), and of dimension $2(c_1(V) \cdot A + n)$.

This statement is no longer true for pseudoholomorphic curves, that is, solutions of (5.11) with $\nu = 0$, as the case of a manifold with $c_1 = 0$ shows.

In this case, the virtual dimension of $W_{J,0,A}$ is equal to $2n$, which is independent of A; but, starting from a pseudoholmorphic curve $f : \mathbb{P}^1 \to V$, we can construct families of pseudoholomorphic curves of arbitrarily large dimension, made up of maps $g : \mathbb{P}^1 \to V$ of the form $g = f \circ \phi$, where $\phi : \mathbb{P}^1 \to \mathbb{P}^1$ is a rational map.

This statement is proved by applying Sard's theorem, showing that for a generic parameter (J,ν) the linearization of Eq. 5.11 at any point (f, J, ν) where 5.11 is satisfied provides a surjective map from the tangent space to the space of f's to the space $\Omega^{0,1}_{\mathbb{P}^1} \otimes f^* T^{1,0}_{V,J}$, which is enough to imply the result by an index computation.

For every integer k we have an *evaluation map*

(5.12)
$$\begin{cases} \mathrm{ev}_k : W_{J,\nu,A} \times (\mathbb{P}^1)^k \to V^k, \\ (f, x_1, \ldots, x_k) \mapsto (f(x_1), \ldots, f(x_k)). \end{cases}$$

We then have (see [**76**], [**81**], [**82**], [**83**]):

THEOREM 5.1. *If V is weakly monotone and J and ν are chosen generically, the boundary of the image of ev_k is of Hausdorff dimension less than or equal to*

$$2(c_1(V) \cdot A + n) + 2k - 2.$$

Similarly, if $(x_1, \ldots, x_l)$ is chosen generically in $(\mathbb{P}^1)^l$, the image under ev_k of $W_{J,\nu,A} \times (x_1, \ldots, x_l) \times (\mathbb{P}^1)^{k-l}$ in V^k has a boundary whose Hausdorff dimension is at most

$$2(c_1(V) \cdot A + n) + 2(k - l) - 2.$$

The invariants $\Phi_A(\alpha_1, \ldots, \alpha_k | \beta_1, \ldots, \beta_l)$ are then multilinear invariants defined as follows.

Let $x_1, \ldots, x_k$ be points chosen generically in $\mathbb{P}^1$. The classes α_i and β_j are classes in $H^*(V)$ modulo torsion. Replacing them by multiples $n_i \alpha_i$ and $m_j \beta_j$ if necessary, we can assume that α_i and β_j are cohomology classes of oriented submanifolds H_{α_i} and H_{β_j} of V.

By Theorem 5.1 and Sard's theorem, we can assume that for

$$\sum_i \deg \alpha_i + \sum_j \deg \beta_j = 2(l + \langle c_1(V), A \rangle + n),$$

the product manifold $H_{\alpha_1} \times \cdots \times H_{\alpha_k} \times H_{\beta_1} \times \cdots \times H_{\beta_l} \subset V^{k+l}$

- does not intersect the boundary of $\mathrm{ev}_{k+l}(W_{J,\nu,A} \times (x_1, \ldots, x_k) \times (\mathbb{P}^1)^l)$ and
- intersects $\mathrm{ev}_{k+l}(W_{J,\nu,A} \times (x_1, \ldots, x_k) \times (\mathbb{P}^1)^l)$ transversally in a finite number of points that can be counted algebraically, since $W_{J,\nu,A} \times (x_1, \ldots, x_k) \times (\mathbb{P}^1)^l$ is a smooth oriented manifold of dimension $2(c_1(V) \cdot A + n) + 2l$.

  The fact that the boundary of $\mathrm{ev}_{k+l}(W_{J,\nu,A} \times (x_1, \ldots, x_k) \times (\mathbb{P}^1)^l)$ is of dimension at most $2(c_1(V) \cdot A + n) + 2l - 2$ implies that the number of these points of intersection, counted according to sign, is independent of the choice of the manifolds H_{α_i} and H_{β_j}.

  To get the product manifold $H_{\alpha_1} \times \cdots \times H_{\alpha_k} \times H_{\beta_1} \times \cdots H_{\beta_l} \subset V^{k+l}$, we then define $\Phi_A(\alpha_1, \ldots, \alpha_k | \beta_1, \ldots, \beta_l)$ as follows:

- when $\sum_i \deg \alpha_i + \sum_j \deg \beta_j = 2(l + \langle c_1(V), A \rangle + n)$ we set

(5.13) $\Phi_A(\alpha_1, \ldots, \alpha_k | \beta_1, \ldots, \beta_l)$
$$= \#\mathrm{ev}_k\big(W_{J,\nu,A} \times (x_1, \ldots, x_k) \times (\mathbb{P}^1)^l\big)$$
$$\cap H_{\alpha_1} \times \cdots \times H_{\alpha_k} \times H_{\beta_1} \times \cdots \times H_{\beta_l},$$

(where the intersection number # is counted with signs defined by the orientations);

- when $\sum_i \deg \alpha_i + \sum_j \deg \beta_j \neq 2(l + \langle c_1(V), A \rangle + n)$, we set

$$\Phi_A(\alpha_1, \ldots, \alpha_k | \beta_1, \ldots, \beta_l) = 0.$$

By using a generic homotopy between two generic couples (J, ν), one can show that this number is independent of the choice of (J, ν). Similarly, this number is independent of the generic choice of $(x_1, \ldots, x_k) \in (\mathbb{P}^1)^k$.

The proof of the property (5.10) then amounts to letting $(\mathbb{P}^1, x_1, \ldots, x_k)$ degenerate over the union of two curves $\mathbb{P}^1_1$ and $\mathbb{P}^1_2$ isomorphic to $\mathbb{P}^1$ intersecting transversally in only one point, with s marked points x_i on the first component forming a generic $(k+1)$-tuple with the point of intersection $x_{s+1} \in \mathbb{P}^1_1$ of the two components and $(k-s)$ marked points x'_i on the second component forming a generic $(k-s+1)$-tuple with the point of intersection $x'_{k-s+1} \in \mathbb{P}^1_2$. Ruan and Tian show that $W_{J,\nu,A}$ then tends to the union over $A_1 + A_2 = A$ of the subspaces of $W_{J,\nu,A_1} \times W_{J,\nu,A_2}$ formed of couples (f_1, f_2) such that

$$f_1(x_{s+1}) = f_2(x'_{k-s+1}).$$

The product $W_{J,\nu,A} \times (x_1, \ldots, x_k) \times (\mathbb{P}^1)^l$ then tends toward the union of the sets

$$W'_{A_1,A_2,i_1,\ldots,i_r} = \{(f_1, f_2, y_1, \ldots, y_l); y_{i_m} \in \mathbb{P}^1_1, y_{j_m} \in \mathbb{P}^1_2,$$
$$f_{i*}([\mathbb{P}^1_i]) = A_i, f_1(x_{s+1}) = f_2(x'_{k-s+1})\}$$

over $A_1 + A_2 = A$ and all partitions $i_1 < \cdots < i_r$ and $j_1 < \cdots < j_{l-r}$ of $\{1, \ldots, l\}$.

For A_1, A_2 and $i_1 < \cdots < i_r$ fixed, we consider the evaluation map with values in V^{k+l+2}:

$$\mathrm{ev}_{i_1,\ldots,i_r} : W_{J,\nu,A_1} \times W_{J,\nu,A_2} \times (\mathbb{P}^1_1)^r \times (\mathbb{P}^1_2)^{l-r}$$

(5.14)
$$\longrightarrow V^{s+1} \times V^{k-s+1} \times V^r \times V^{l-r},$$
$$(f_1, f_2, y, z) \mapsto (f_1(x), f_2(x'), f_1(y), f_2(z))$$

where $x = (x_1, \ldots, x_{s+1})$ and $x' = (x'_1, \ldots, x'_{k-s+1})$.

Let $\Delta \subset V^{k+l+2}$ be the submanifold defined by the condition $v_{s_1} = v_{k+2}$. From the preceding, the intersection

$$\mathrm{ev}_{k+l}\big(W_{J,\nu,A} \times (x_1, \ldots, x_k) \times (\mathbb{P}^1)^l\big) \cap (H_{\alpha_1} \times \cdots H_{\alpha_k} \times H_{\beta_1} \times \cdots \times H_{\beta_l})$$

converges to

(5.15) $$\bigcup_{\substack{A_1+A_2=A \\ i_1<\cdots<i_r}} \mathrm{ev}_{i_1,\ldots,i_r}\big(W_{J,\nu,A_1} \times W_{J,\nu,A_2} \times (\mathbb{P}^1_1)^r \times (\mathbb{P}^1_2)^{l-r}\big)$$

$$\cap \pi^{-1}(H_{\alpha_1} \times \cdots \times H_{\alpha_k} \times H_{\beta_1} \times \cdots \times H_{\beta_l}) \cap \Delta),$$

where $\pi : V^{k+l+2} \to V^{k+l}$ is the projection that forgets points $(s+1)$ and $(k+2)$. Since the diagonal of $V \times V$ is homologous to a combination $\sum_{\sigma,\tau} g^{\sigma\tau} H_\sigma \times H_\tau$ for which the homology classes of the H_σ form a basis of $H_*(V,\mathbb{Q})$ and where $g^{\sigma\tau}$ is the inverse of the intersection matrix $\langle H_\sigma, H_\tau\rangle$, we find

$$(5.16) \quad \#\mathrm{ev}_{i_1,\ldots,i_r}\bigl(W_{J,\nu,A_1}\times W_{J,\nu,A_2}\times(\mathbb{P}_1^1)^r\times(\mathbb{P}_2^1)^{k-r}\bigr)$$
$$\cap \pi^{-1} H_{\alpha_1}\times\cdots\times H_{\alpha_k}\times H_{\beta_1}\times\cdots\times H_{\beta_l}\cap\Delta$$
$$=\sum_{\sigma,\tau} g^{\sigma\tau}\#\mathrm{ev}_{i_1,\ldots,i_r}\bigl(W_{J,\nu,A_1}\times W_{J,\nu,A_2}\times(\mathbb{P}_1^1)^r\times(\mathbb{P}_2^1)^{k-r}\bigr)$$
$$\cap H_{\alpha_1}\times\cdots\times H_{\alpha_s}\times H_\sigma\times\cdots H_{\alpha_{s+1}}$$
$$\times\cdots\times H_{\alpha_k}\times H_\tau\times H_{\beta_1}\times\cdots\times H_{\beta_l}$$

which by definition equals

$$(5.17) \quad \sum_{\sigma,\tau} g^{\sigma\tau}\Phi_A(\alpha_1,\ldots,\alpha_s,e_\sigma|\beta_{i_1},\ldots,\beta_{i_r})\Phi_{A_2}(\alpha_{s+1},\ldots,\alpha_k,e_\tau|\beta_{j_1},\ldots,\beta_{j_{l-r}})$$

where e_σ is the image of the class of H_σ under Poincaré duality. This proves formula (5.10), after a rigorous analysis of the convergence that forms the major content of [83]. $\square$

Another important property of these mixed invariants is their (anti)symmetry relative to the classes α_i and β_j.

PROPOSITION 5.2 (see [83]). *The following relations hold:*

$$\Phi_A(\alpha_1,\ldots,\alpha_k|\beta_1,\ldots,\beta_l)$$
$$=(-1)^{\deg\alpha_i\deg\alpha_{i+1}}\Phi_A(\alpha_1,\ldots,\alpha_{i+1},\alpha_i,\ldots,\alpha_k|\beta_1,\ldots,\beta_l),$$

$$\Phi_A(\alpha_1,\ldots,\alpha_k|\beta_1,\ldots,\beta_l)$$
$$=(-1)^{\deg\beta_i\deg\beta_{i+1}}\Phi_A(\alpha_1,\ldots,\alpha_k|\beta_1,\ldots,\beta_{i+1},\beta_i,\ldots,\beta_l),$$

The proof is immediate. The second assertion follows for example, from the fact that if σ is the involution that interchanges the factors of order $k+i$ and $k+i+1$ on V^{k+l}, we have

$$(5.18)\quad \sigma^*(\alpha_1\otimes\cdots\otimes\alpha_k\otimes\beta_1\otimes\cdots\otimes\beta_l)$$
$$=(-1)^{\deg\beta_i\deg\beta_{i+1}}(\alpha_1\otimes\cdots\otimes\alpha_k\otimes\beta_1\otimes\cdots\otimes\beta_{i+1}\otimes\beta_i\otimes\cdots\otimes\beta_l)$$

while σ changes neither the orientation of V^{k+l} nor

$$\mathrm{ev}_{k+l}\bigl(W_{A,J,\nu}\times(x_1,\ldots,x_k)\times(\mathbb{P})^l\bigr). \quad \square$$

2.3. Connection with the Kontsevich–Manin invariants. Between the mixed invariants $\Phi_A(\alpha_1,\ldots,\alpha_k|\beta_1,\ldots,\beta_l)$ and the invariants $H_{A,g,k}$ of 1.1 we should have the following connection:

(5.19)
$$\Phi_A(\alpha_1,\alpha_2,\alpha_3|\beta_1,\ldots,\beta_{k-3})=\int_{\overline{\mathcal{M}}_{0,k}} H_{A,0,k}(\alpha_1\otimes\alpha_2\otimes\alpha_3|\beta_1\otimes\cdots\otimes\beta_{k-3}).$$

Indeed, if one could assume $\nu=0$, the space "$\mathrm{Mor}_A(0,V,k)$" would be the quotient of $W_{J,\nu,A}\times(\mathbb{P}^1)^k$ by $\mathrm{Aut}\,\mathbb{P}^1$, and we would have the following commutative diagram:

$$W_{J,\nu,A} \times x_1 \times x_2 \times x_3 \times (\mathbb{P}^1)^{k-3} \xrightarrow{j} W_{J,\nu,A} \times (\mathbb{P}^1)^k$$

(5.20)

$$\phi \searrow \quad \downarrow \quad \searrow \mathrm{ev}_k$$

$$\mathrm{Mor}_A(0,V,k) \xrightarrow{\mathrm{ev}} V^k.$$

The assertion then follows from the fact that ϕ is birational, so that the integral over $W_{J,\nu,A} \times x_1 \times x_2 \times x_3 \times (\mathbb{P}^1)^{k-3}$ of

$$\mathrm{ev}_k^*(\alpha_1 \otimes \alpha_2 \otimes \alpha_3 \otimes \beta_1 \otimes \cdots \otimes \beta_{k-3})$$

(that is, $\Phi_A(\alpha_1,\alpha_2,\alpha_3|\beta_1,\ldots,\beta_{k-3})$) is equal to the integral over $\mathrm{Mor}_A(0,V,k)$ of

$$\mathrm{ev}^*(\alpha_1 \otimes \alpha_2 \otimes \alpha_3 \otimes \beta_1 \otimes \cdots \otimes \beta_{k-3})$$

(that is, $\int_{\overline{\mathcal{M}}_{0,k}} H_{A,0,k}(\alpha_1 \otimes \alpha_2 \otimes \alpha_3|\beta_1 \otimes \cdots \otimes \beta_{k-3})$).

More generally, $\Phi_A(\alpha_1,\ldots,\alpha_k|\beta_1,\ldots,\beta_l)$ should be obtained as the integral of $H_{a,0,k+l}(\alpha_1 \otimes \cdots \otimes \alpha_k \otimes \beta_1 \otimes \cdots \otimes \beta_l)$ on the submanifold of $\overline{\mathcal{M}}_{0,k+l}$ that is the closure of the image of $(x_1,\ldots,x_k) \times (\mathbb{P})^l$.

It would be interesting to know if the mixed invariants are sufficient to determine the whole data of $H_{A,0,k}$.

On the other hand, the connection between formula (5.10) and Axiom A6 of 1.2 is the following. Fix four distinct points t_i of $\mathbb{P}^1$. These points determine a divisor D_t of $\overline{\mathcal{M}}_{0,4+l}$ formed from the $(4+l)$-tuplets

$$z = (z_1,\ldots,z_{4+l}) \text{ such that } z \equiv (t_1,\ldots,t_4,z_1',\ldots,z_l') \text{ modulo } \mathrm{Aut}\,\mathbb{P}^1.$$

But we have the following proposition.

LEMMA 5.3. *The divisor D_t is numerically equal to*

$$\sum_{\substack{i_1 < \cdots < i_r \\ r \leq l}} \phi_I(\overline{\mathcal{M}}_{0,3+r} \times \overline{\mathcal{M}}_{0,3+l-r})$$

where $\phi_I((x_1,\ldots,x_{3+r}),(y_1,\ldots,y_{3+l-r}))$ is equal to $\left(\mathbb{P}^1 \bigcup_{x_3=y_3} \mathbb{P}^1, z_1,\ldots,z_{4+l}\right)$, with $z_1 = x_1$, $z_2 = x_2$, $z_{i_k} = x_{2+k}$, $z_3 = y_1$, $z_4 = y_2$, and $z_{j_k} = y_k$.

Indeed, this can be seen by letting $(\mathbb{P}^1,t_1,\ldots,t_4)$ become degenerate on the set $\left(\mathbb{P}_1^1 \bigcup_{x=y} \mathbb{P}_2^1, t_1,\ldots,t_4\right)$, with $t_1,t_2, \in \mathbb{P}_1^1$ and $t_3,t_4 \in \mathbb{P}_2^1$. □

Just as the invariant $\Phi_A(\alpha_1,\ldots,\alpha_4|\beta_1,\ldots,\beta_l)$ should be interpreted as the integral $\int_{D_t} H_{A,0,4+l}(\alpha_1 \otimes \cdots \otimes \alpha_4 \otimes \beta_1 \otimes \cdots \otimes \beta_l)$ in the Kontsevich–Manin formulation, we see that, for $k = 4$, the formula (5.10) is the integrated version of Axiom A6.

3. Gromov–Witten potential

Let us fix a symplectic form ω on V. The Gromov–Witten potential (see [78]) is then a function on $H^{2*}(V,\mathbb{C})$ defined, under suitable convergence hypotheses, by

(5.21) $$\Phi_\omega(\alpha) = \sum_{\substack{k \geq 3 \\ A \in H_2(V,\mathbb{Z})}} \frac{1}{k!} \exp\left(-\int_A \omega\right) \int_{\overline{\mathcal{M}}_{0,k}} H_{A,0,k}(\alpha \otimes \cdots \otimes \alpha).$$

(Here we are following the terminology of Kontsevich–Manin, but we recall that the invariants $\int_{\overline{M}_{0,k}} H_{A,0,k}(\alpha \otimes \cdots \otimes \alpha)$ are well defined by the work of Ruan and Tian (see 2).)

There are no general results on the convergence of this expression in a non-empty open set of $H^{2*}(V,\mathbb{C})$. However, if V is a Fano variety (or in the symplectic context, if $c_1(-K)$ is a positive multiple of the class of ω), for k fixed, there is only a finite number of $A \in H_2(V,\mathbb{Z})$ such that $\int_{\overline{M}_{0,k}} H_{A,0,k}(\alpha \otimes \cdots \otimes \alpha) \neq 0$, by compactness [76] and for reasons of dimension. On the other hand, again by compactness, for every constant $C > 0$ there exists only one finite number of $A \in H_2(V,\mathbb{Z})$ such that $\int_A \omega \leq C$ and which occurs in the expression for Φ_ω.

3.1. The case of Calabi–Yau threefolds. When the canonical bundle of V is trivial, one can see immediately that the components of the space of holomorphic maps of $\mathbb{P}^1$ into V are of virtual complex dimension 3, and hence are virtually made up of constants in the case $A = 0$ and of reparametrizations of a rigid rational curve on V. (Actually one should always keep in mind the branched coverings of a given curve that give families of dimension strictly larger than 3. We have explained in § 2 how this difficulty was overcome by Ruan and Tian, and we shall explain in § 6 the calculation of the contributions of these "excess" components to the Gromov–Witten invariants.)

By Axioms A1–A6, or by their partial justification as given by Ruan and Tian, we can easily derive the following result.

PROPOSITION 5.4. *If V is a Calabi–Yau threefold, the Gromov–Witten potential Φ_ω of V has the following form, modulo a quadratic function in α:*

$$\Phi_\omega(\alpha) = \frac{1}{6} \int_V \alpha^3 + \sum_{A \neq 0} N(A) \exp\left(\int_A -\omega + \alpha \right), \tag{5.22}$$

where $N(A)$ is the virtual number of rational curves of class A in V.

Indeed, among the terms corresponding to $A = 0$, the only nonzero term comes from the case in which $\dim \overline{M}_{0,k} = 0$ (by Axiom A5), and is equal (by the same axiom) to $\frac{1}{6} \int_V \alpha^3$.

On the other hand, if $A \neq 0$ and $k \geq 3$, from the fact that the rational nonconstant curves are virtually rigid, we find immediately

$$\int_{\overline{M}_{0,k}} H_{A,0,k}(\alpha_1 \otimes \cdots \otimes \alpha_k) = 0$$

if $\deg \alpha_i > 2$ for at least one index i.

If on the other hand $k \geq 4$ and $\deg \alpha_k = 2$, this term equals

$$\int_A \alpha_k \int_{\overline{M}_{0,k-1}} H_{A,0,k-1}(\alpha_1 \otimes \cdots \otimes \alpha_{k-1})$$

by Axiom A4, while if $\deg \alpha_k = 0$, it is zero by Axiom A3.

Finally, for $k = 3$ and $A \neq 0$ the equality

$$\int_{\overline{M}_{0,3}} H_{A,0,3}(\alpha_1 \otimes \alpha_2 \otimes \alpha_3) = N(A) \int_A \alpha_1 \int_A \alpha_2 \int_A \alpha_3$$

is clear by the virtual definition of $H_{A,0,k}$.

We thus find that the contribution of the sum for fixed $A \neq 0$ to Φ_ω is, modulo a quadratic term in α, equal to

$$(5.23) \qquad \sum_{k \geq 3} \frac{1}{k!} N(A) \exp\left(-\int_A \omega\right)\left(\int_A \alpha\right)^k \equiv N(A) \exp\left(\int_A -\omega + \alpha\right). \qquad \square$$

3.2. The WDVV equation. Here we are following [75]. Let $t_1, \ldots, t_n$ be coordinates on a manifold M and (g_{ij}) a nondegenerate symmetric matrix with constant coefficients defining an inner product $\langle , \rangle$ on T_M. (We admit the case of holomorphic coordinates, the inner product then being $\mathbb{C}$-bilinear on the holomorphic tangent bundle.) Let $f(t)$ be a function of class $\mathcal{C}^3$ (holomorphic in the holomorphic case) satisfying the condition

$$(5.24) \qquad \frac{\partial^3 f}{\partial t_1 \partial t_i \partial t_j} = g_{ij}.$$

We then construct at every point $t \in M$ a commutative unitary product "$\bullet_t$" on $T_{M,t}$ defined by the condition

$$(5.25) \qquad \left\langle \frac{\partial}{\partial t_i} \bullet_t \frac{\partial}{\partial t_j}, \frac{\partial}{\partial t_k} \right\rangle = \frac{\partial^3 f}{\partial t_i \partial t_j \partial t_k}(t).$$

Clearly $\partial/\partial t_1$ is the unit element by (5.24) and (5.25).

The WDVV (Witten–Dijgraaf–Verlinde–Verlinde) equation expresses the condition that the product "$\bullet_t$" be associative at every point. By commutativity, the condition of associativity is equivalent to symmetry of the expression

$$\left\langle \left(\frac{\partial}{\partial t_i} \bullet_t \frac{\partial}{\partial t_j}\right) \bullet_t \frac{\partial}{\partial t_k}, \frac{\partial}{\partial t_l} \right\rangle$$

in the indices j and k for all i, l, j, k. Setting

$$C_{ijk} = \frac{\partial^3 f}{\partial t_i \partial t_j \partial t_k},$$

we have by definition

$$\frac{\partial}{\partial t_i} \bullet_t \frac{\partial}{\partial t_j} = \sum_{a,b} g^{ab} C_{ija} \frac{\partial}{\partial t_b},$$

so that the WDVV equation can be written

$$(5.26) \qquad \sum_{a,b} C_{ija} g^{ab} C_{klb} = \sum_{a,b} C_{ika} g^{ab} C_{jlb} \quad \forall i, l, j, k.$$

3.3. The connection. Let ∇ be the connection on T_M for which the $\partial/\partial t_i$ are parallel (that is, the Levi-Cività connection of g); using the product "$\bullet_t$," we can construct for every $z \in \mathbb{R}$ (or $\mathbb{C}$ in the holomorphic case) a torsion-free connection on T_M by means of the formula

$$(5.27) \qquad \widetilde{\nabla}^z_u(v) = \nabla_u(v) + zu \bullet_t v$$

for u and v vector fields on M. We then have the following proposition (see [75]).

PROPOSITION 5.5. *If f satisfies the WDVV equation, the connection $\widetilde{\nabla}^z$ is flat for all z.*

Indeed, we can verify immediately that the vanishing of the term in z^2 on the curvature $\left(\widetilde{\nabla}^z\right)^2$ is equivalent to the associativity of the product "$\bullet_t$," while the symmetry of the derivatives $\frac{\partial}{\partial t_l}C_{ijk} = \frac{\partial}{\partial t_k}C_{ijl}$ assures the vanishing of the term in z. $\square$

Conversely, if we have a product "$\bullet_t$," described as above by the coefficients $C_{ijk}(t)$, assumed symmetric in i, j, k, such that the connection $\widetilde{\nabla}^z$ is flat for all t, then the product is associative by the vanishing of the term in z^2 on the curvature $(\widetilde{\nabla}^z)^2$, and the vanishing of the term in z implies that there exists a function f such that $C_{ijk} = \partial^3 f/\partial t_i \partial t_j \partial t_k$.

Returning to the Gromov–Witten potential, which is a function on the vector space $H^{2*}(V,\mathbb{C})$, we then have the following result (see [78], [83]).

PROPOSITION 5.6. *The Gromov–Witten potential Φ_ω (assumed convergent) satisfies the WDVV equation for the natural flat structure of $H^{2*}(V,\mathbb{C})$, the metric being given by the intersection form, and the unit vector field being equal to 1_V.*

FIRST PROOF. We begin with the proof of Kontsevich and Manin, even though it uses Axioms A1–A6, which are not demonstrated in that generality by Ruan and Tian. We shall show later that the property (5.10) of the invariants of Ruan and Tian suffices to imply Proposition 5.6.

It is first necessary to verify that we do indeed have, for the linear coordinates associated with a basis e_i of $H^{2*}(V,\mathbb{C})$ such that $e_1 = 1_V$, the relation

$$(5.28) \qquad \frac{\partial^3 \Phi_\omega}{\partial t_1 \partial t_i \partial_j} = g_{ij} = \int_V e_i \wedge e_j.$$

But this follows immediately from Axioms A2 and A3. We find

$$\frac{\partial^3 \Phi_\omega}{\partial t_1 \partial t_i \partial t_j} = \sum_{\substack{k \geq 3 \\ A \in H_2(V,\mathbb{Z})}} \frac{e^{-\int_A \omega}}{(k-3)!} \int_{\overline{M}_{0,k}} H_{A,0,k}\left(1_V \otimes e_i \otimes e_j \otimes \alpha^{\otimes k-3}\right).$$

By Axiom A3, the only nonzero term corresponds to the case $k = 3$ and $A = 0$ (since for $k > 3$ the classes $H_{A,0,k}(1_V \otimes \alpha_1 \otimes \cdots \otimes \alpha_{k-1})$ are the inverse images of classes on $\overline{M}_{0,k-1}$ and hence have integral zero), and equals $\int_V e_i \wedge e_j$ (by the second part of Axiom A3).

It remains to verify Eq. (5.26), that is, for all i, l, j, k the expression

$$\sum_{a,b} \frac{\partial^3 \Phi_\omega}{\partial t_i \partial t_j \partial t_a} g^{ab} \frac{\partial^3 \Phi_\omega}{\partial t_k \partial t_l \partial t_b}$$

is symmetric in j and k. But we have

$$(5.29) \qquad \frac{\partial^3 \Phi_\omega}{\partial t_i \partial t_j \partial t_a}(\alpha) = \sum_{\substack{r \geq 3 \\ A \in H_2(V,\mathbb{Z})}} \frac{e^{-\int_A \omega}}{(r-3)!} \int_{\overline{M}_{0,r}} H_{A,0,k}(e_i \otimes e_j \otimes e_a \otimes \alpha^{\otimes r-3}).$$

3. GROMOV–WITTEN POTENTIAL

It thus follows that

$$(5.30) \quad \sum_{a,b} \frac{\partial^3 \Phi_\omega}{\partial t_i \partial t_j \partial t_a} g^{ab} \frac{\partial^3 \Phi_\omega}{\partial t_k \partial t_l \partial t_b}$$

$$= \sum_{a,b} \sum_{\substack{r,s \geq 3 \\ A_1, A_2}} g^{ab} e^{-\int_{A_1+A_2} \omega} \frac{1}{(r-3)!(s-3)!}$$

$$\times \int_{\overline{\mathcal{M}}_{0,r}} H_{A_1,0,r}(e_i \otimes e_j \otimes e_a \otimes \alpha^{\otimes r-3})$$

$$\times \int_{\overline{\mathcal{M}}_{0,s}} H_{A_2,0,s}(e_k \otimes e_l \otimes e_b \otimes \alpha^{\otimes s-3}).$$

By Axiom A6 with $g_1 = g_2 = 0$, the term that appears on the right when we carry out the summation for $A = A_1 + A_2$ and r and s fixed is equal to

$$\frac{1}{(r-3)!(s-3)!} \exp\left(-\int_A \omega\right) \int_{D_r} H_{A,0,n}(\alpha^{\otimes r-3} \otimes e_i \otimes e_j \otimes \alpha^{\otimes s-3} \otimes e_k \otimes e_l),$$

where $n = r + s - 2$ and D_r is the divisor $\phi(\overline{\mathcal{M}}_{0,r} \times \overline{\mathcal{M}}_{0,s})$ of $\overline{\mathcal{M}}_{0,n}$ (see 1.2).

Indeed, it is necessary only to note that Axiom A6 remains true if we confine ourselves to the even cohomology, the sum extending over σ and τ such that e_σ is a basis of $H^{2*}(V)$, from the fact that $\overline{\mathcal{M}}_{0,n}$ has no cohomology of odd degree. The symmetry of this expression in j and k then results from the relations due to Keel [77] among the divisors of $\overline{\mathcal{M}}_{0,n}$.

Indeed, for any partition $S = S_1 \sqcup S_2$ of $\{1, \ldots, n\}$ such that $|S_1| \geq 2$ and $|S_2| \geq 2$, we have a divisor D_S of $\overline{\mathcal{M}}_{0,n}$ formed of reducible curves with n marked points such that the points indexed by S_1 are on one of the components and those indexed by S_2 are on the other. We then have the following.

PROPOSITION 5.7 (see [77]). *Let $i, j, k, l \in \{1, \ldots, n\}$. We then have the relation*

$$(5.31) \quad \sum_{\substack{i,j \in S_1 \\ k,l \in S_2}} D_S = \sum_{\substack{i,k \in S_1 \\ j,l \in S_2}} D_S.$$

PROOF. We may certainly assume that $\{i, j, k, l\} = \{1, 2, 3, 4\}$, by letting the symmetric group act on $\overline{\mathcal{M}}_{0,n}$ if necessary. We then use—as in the proof of Lemma 5.3, whose notation we adopt—the divisors D_t, with $t \in \overline{\mathcal{M}}_{0,4}$ which are all homologous in $\overline{\mathcal{M}}_{0,n}$.

The proof of Proposition 5.7 is carried out by letting $t = (\mathbb{P}^1, t_1, \ldots, t_4)$ degenerate on $(\mathbb{P}^1_1 \underset{x=y}{\bigcup} \mathbb{P}^1_2, t_1, \ldots, t_4)$ with $t_1, t_3 \in \mathbb{P}^1_1$ and $t_2, t_4 \in \mathbb{P}^1_2$. (In the proof of Lemma 5.3 we let t degenerate on $(\mathbb{P}^1_1 \underset{x=y}{\bigcup} \mathbb{P}^1_2, t_1, \ldots, t_4)$ with $t_1, t_2 \in \mathbb{P}^1_1$ and $t_3, t_4 \in \mathbb{P}^1_2$.)

We then find that D_t is homologous to

$$\sum_{\substack{i_1 < \cdots < i_r \\ r \leq l}} \phi_{I'}(\overline{\mathcal{M}}_{0,3+r} \times \overline{\mathcal{M}}_{0,3+l-r}),$$

where $\phi_{I'}\big((x_1,\ldots,x_{3+r}),(y_1,\ldots,y_{3+l-r})\big)$ is equal to $\Big(\mathbb{P}^1 \underset{x_3=y_3}{\bigcup} \mathbb{P}^1, z_1,\ldots,z_{4+l}\Big)$ with $z_1 = x_1$, $z_2 = y_1$, $z_{i_k} = x_{2+k}$, $z_3 = x_2$, $z_4 = y_4$, and $z_{j_k} = y_k$, which, combined with Lemma 5.3, proves Proposition 5.7. $\square$

This implies the result, since we then find that for n fixed

$$(n-4)! \sum_{r+s=n+2} \frac{e^{-\int_A \omega}}{(r-3)!(s-3)!} \int_{D_r} H_{A,0,n}(\alpha^{\otimes r-3} \otimes e_i \otimes e_j \otimes \alpha^{\otimes s-3} \otimes e_k \otimes e_l)$$

$$= \sum_{\substack{r-2,r-1 \in S_1 \\ n-1, n \in S_2}} \int_{D_S} H_{A,0,n}(\alpha^{\otimes r-3} \otimes e_i \otimes e_j \otimes \alpha^{\otimes s-3} \otimes e_k \otimes e_l)$$

is symmetric in j and k. $\square$

SECOND PROOF. We now show Proposition 5.6 by using the results of Ruan and Tian, and particularly the formula (5.10). In the terminology of [**83**], this proposition asserts that the function

$$(5.32) \qquad \Phi_\omega(\alpha) = \sum_{k \geq 3, A} \frac{e^{-\int_A \omega}}{k!} \Phi_A\big(\alpha,\alpha,\alpha \big| \underbrace{\alpha,\ldots,\alpha}_{k-3}\big)$$

satisfies the WDVV equation (5.26), which means that the expression

$$(5.33) \qquad \sum_{a,b} g^{ab} \partial^3_{ija} \Phi_\omega \partial^3_{klb} \Phi_\omega$$

is symmetric in the indices j and k, the partial derivatives being taken with respect to the linear coordinates on $H^{2*}(V)$ associated with a basis e_a. In fact we have

$$(5.34) \qquad \partial^3_{ijk} \Phi_\omega(\alpha) = \sum_{k \geq 3, A} \frac{e^{-\int_A \omega}}{(k-3)!} \Phi_A\big(e_i, e_j, e_k \big| \underbrace{\alpha,\ldots,\alpha}_{k-3}\big).$$

This follows from the multilinearity of $\Phi_A(\alpha_1,\alpha_2,\alpha_3|\alpha_4,\ldots,\alpha_k)$ in $\alpha_i \in H^{2*}(V,\mathbb{C})$ for $i = 1,\ldots,k$ and the following fact:

The mixed invariants $\Phi_A(\alpha_1,\alpha_2,\alpha_3|\alpha_4,\ldots,\alpha_k)$ are symmetric in the α_i in $H^{2*}(V,\mathbb{C})$ for $i = 1,\ldots,k$.

The expression (5.33) thus becomes

$$(5.35) \qquad \sum_{\substack{k_1,k_2 \geq 3 \\ A_1,A_2}} \sum_{a,b} g^{ab} \frac{1}{(k_1-3)!(k_2-3)!} \exp\Big(-\int_{A_1+A_2} \omega\Big)$$

$$\Phi_{A_1}\big(e_i, e_j, e_a \big| \underbrace{\alpha,\ldots,\alpha}_{k_1-3}\big) \Phi_{A_2}\big(e_k, e_l, e_b \big| \underbrace{\alpha,\ldots,\alpha}_{k_2-3}\big).$$

But, by (5.10) we have for $k \geq 0$ and A fixed,

$$(5.36) \quad \sum_{\substack{A_1+A_2=A \\ a,b}} g^{ab} \exp\left(-\int_A \omega\right) \sum_{k_1+k_2-6=k} \frac{1}{(k_1-3)!(k_2-3)!}$$

$$\Phi_{A_1}\big(e_i, e_j, e_a \big| \underbrace{\alpha, \ldots, \alpha}_{k_1-3}\big) \Phi_{A_2}\big(e_k, e_l, e_b \big| \underbrace{\alpha, \ldots, \alpha}_{k_2-3}\big)$$

$$= \frac{1}{k!} \Phi_A(e_i, e_j, e_k, e_l | \underbrace{\alpha, \ldots, \alpha}_{k})$$

and the last term is symmetric in j and k by Proposition 5.2. □

4. Application to mirror symmetry

In the case of Calabi–Yau threefolds, the WDVV equation can be verified immediately because of the particular form (5.22) of the Gromov–Witten potential. (Except for the cubic term, $\Phi_\omega(\alpha)$ depends only on the component of α in $H^2(V, \mathbb{C})$, which is a totally isotropic subspace of $H^{2*}(V, \mathbb{C})$ for the metric g_{ij}.) The Gromov–Witten potential (assuming convergence) makes it possible in this case to construct a variation of complex Hodge structure parameterized by $H^2(V, \mathbb{C})$ (which should be that of the mirror) as follows. We have first of all,

LEMMA 5.8. *If V is a Calabi–Yau threefold, then for all α in $H^{2*}(V, \mathbb{C})$ such that Φ_ω converges in a neighborhood of α, the product "$\bullet_\alpha$" on $H^{2*}(V, \mathbb{C})$, constructed using the cubic derivatives of Φ_ω and the metric g_{ij}, as in (5.25), preserves the grading of $H^{2*}(V, \mathbb{C})$.*

PROOF. We recall that we have

$$\Phi_\omega(\alpha) = \frac{1}{6}\int_V \alpha^3 + \sum_{A \neq 0} N(A) \exp\left(\int_A -\omega + \alpha\right)$$

modulo a quadratic term in α. From that we deduce that if e_i is a basis of $H^{2*}(V, \mathbb{C})$ made of homogeneous elements and t_i are the corresponding coordinates, we have

$$\frac{\partial^3 \Phi_\omega}{\partial t_i \partial t_j \partial t_k}(\alpha) = \int_V e_i \wedge e_j \wedge e_k$$

if one of the e_l is not of degree 2. From that we deduce by definition of the product "$\bullet_\alpha$" that

$$\langle e_i \bullet_\alpha e_j, e_k \rangle = \int_V e_i \wedge e_j \wedge e_k$$

if one of the e_l is not of degree 2. Consequently

- if e_i or e_j is not of degree 2, we have $e_i \bullet_\alpha e_j = e_i \wedge e_j$;
- otherwise $e_i \bullet_\alpha e_j$ differs from $e_i \wedge e_j$ by an element orthogonal to $\bigoplus_{k \neq 1} H^{2k}(V)$, that is, by an element of $H^4(V)$.

The product "$\bullet_\alpha$" thus does indeed preserve the grading. □

4.1. The variation of Hodge structure. Here we follow [**12**]. By Propositions 5.4 and 5.5 Φ_ω makes it possible to construct a flat (holomorphic) connection $\widetilde{\nabla}^1$ on the (holomorphic) tangent bundle of $H^{2*}(V,\mathbb{C})$ by means of the formula

$$\widetilde{\nabla}^1_u(v) = \nabla_u(v) + u \bullet_\alpha v. \tag{5.37}$$

Lemma 5.8 shows that for u tangent to $H^2(V,\mathbb{C})$ and a section v of $\bigoplus_{k \leq r} H^{2k}(V,\mathbb{C})$ we have

$$\widetilde{\nabla}^1_u(v) \in \bigoplus_{k \leq r+1} H^{2k}(V,\mathbb{C}).$$

From that we deduce that the filtration on $H^{2*}(V,\mathbb{C})$ defined by

$$F^3 = H^0, \quad F^2 = H^0 \oplus H^2, \quad F^1 = H^0 \oplus H^2 \oplus H^4, \quad F^0 = H^{2*} \tag{5.38}$$

satisfies the Griffiths transversality condition (Theorem 1.16) for the connection $\widetilde{\nabla}^1$ restricted to $H^2(V,\mathbb{C})$.

We have $\operatorname{rank} F^3 = \operatorname{rank} F^0/F^1 = 1$ and also $\operatorname{rank} F^2/F^3 = \operatorname{rank} F^1/F^2 = \operatorname{rank} H^2(V,\mathbb{C})$. On the other hand, the map of infinitesimal variation of Hodge structure

$$d\mathcal{P} : T_{H^2(V)} \longrightarrow \operatorname{Hom}\left(F^3 H^{2*}(V), F^2 H^{2*}(V)/F^3 H^{2*}(V)\right)$$

defined by $d\mathcal{P}(u)(\omega) = \widetilde{\nabla}^1_u(\omega)$ modulo ω is an isomorphism. Indeed, $F^3 H^{2*}(V)$ is generated by 1_V and by definition of $\widetilde{\nabla}^1$ we have

$$d\mathcal{P}(u)(1_V)_\omega = 1_V \bullet_\omega u = u \in H^2(V) \subset H^{2*}(V) \quad \text{modulo } H^0(V).$$

We have thus constructed a complex variation of Hodge structure having the same numerical characteristics as that of a complete family of Calabi–Yau threefolds. Naturally we expect that this is the variation of Hodge structure of the mirror family.

We note that this construction gives supplementary evidence for mirror symmetry and completes that of 1, where it was shown that there exist special coordinates (assumed to correspond to the flat structure on the space of complexified Kähler parameters of the mirror) defined on the universal covering of the moduli space of a Calabi–Yau threefold and a potential whose third derivatives calculate the normalized Yukawa couplings Y_2.

Here we have taken a step in the opposite direction by constructing a variation of complex Hodge structure with Hodge number $h^{2,1} = h^{1,1}(V)$ parameterized by $H^2(V,\mathbb{C})$ for a Calabi–Yau threefold V, assuming the convergence of the Gromov–Witten potential.

5. Quantum product

Having defined the invariants $\Phi_A(\alpha_1,\ldots,\alpha_k)$ as in (5.13) with $l=0$, we construct the quantum product "$\bullet_\omega$" on $H^*(V)$, where ω is the initial symplectic form on V or a deformation of it or a "complexified Kähler parameter" (see 8) by means of the formula

$$\langle \alpha \bullet_\omega \beta, \gamma \rangle = \sum_{A \in H_2(V,\mathbb{Z})} \Phi_A(\alpha,\beta,\gamma) \exp\left(-\int_A \omega\right), \tag{5.39}$$

for $\alpha, \beta, \gamma \in H^*(V)$, the bilinear form $\langle\,,\,\rangle$ being the intersection form on $H^*(V)$.

We assume here that the term on the right-hand side is a convergent series.

We shall explain in the next chapter how one can give a formal meaning to this product using the Novikov ring of (V,ω). One of the applications of the formula (5.10) is then the following result, which is an immediate generalization of Proposition 5.6.

PROPOSITION 5.9. *The product "$\bullet_\omega$" is associative.*

PROOF. We note that by Proposition 5.2 the product "$\bullet_\omega$" is commutative in the graded sense, that is, it satisfies

$$\alpha \bullet_\omega \beta = (-1)^{\deg(\alpha)\deg\beta} \beta \bullet_\omega \alpha. \tag{5.40}$$

The associativity of "$\bullet_\omega$" on $H^*(V)$ is then equivalent to the property

$$\langle (\alpha_1 \bullet_\omega \alpha_2) \bullet_\omega \alpha_3, \alpha_4 \rangle = (-1)^{\deg\alpha_2 \deg\alpha_3} \langle (\alpha_1 \bullet_\omega \alpha_3) \bullet_\omega \alpha_2, \alpha_4 \rangle \tag{5.41}$$

for homogeneous elements $\alpha_i \in H^*(V)$ and $i = 1,\ldots,4$. But we have by definition

$$\alpha_1 \bullet_\omega \alpha_2 = \sum_{A,\sigma,\tau} g^{\sigma\tau} \Phi_A(\alpha_1,\alpha_2,e_\sigma) \exp\left(-\int_A \omega\right) e_\tau \tag{5.42}$$

where e_σ is a basis of $H^*(V)$ and $g^{\sigma\tau}$ is the inverse of the intersection matrix. It then follows that

$$\langle (\alpha_1 \bullet_\omega \alpha_2) \cdot \alpha_3, \alpha_4 \rangle \tag{5.43}$$
$$= \sum_{A_1,A_2,\sigma,\tau} g^{\sigma\tau} \Phi_{A_2}(e_\tau,\alpha_3,\alpha_4) \Phi_{A_1}(\alpha_1,\alpha_2,e_\sigma) \exp\left(-\int_{A_1+A_2} \omega\right).$$

By formula (5.10) the second member is equal to

$$\sum_A \Phi_a(\alpha_1,\ldots,\alpha_4) \exp\left(-\int_A \omega\right), \tag{5.44}$$

and the (anti)symmetry in α_2 and α_3 then follows from Proposition 5.2. $\square$

REMARK 5.10. We can restrict the quantum product to $H^{2*}(V)$. In that case it follows from the definition that "$\bullet_\omega$" is the product defined on $H^{2*}(V)$ using the cubic derivatives of Φ_ω at 0 as in (5.25). Proposition 5.9 for the even cohomology then follows from the fact that Φ_ω satisfies the WDVV equation.

6. The calculation of Aspinwall and Morrison

We consider a Calabi–Yau threefold. We have explained in the preceding section how to calculate the Gromov–Witten potential using the solutions of the Cauchy–Riemann equation with an inhomogeneous term. However, this potential and its third derivatives (which should give the Yukawa couplings Y_1 of (2.48) and (2.54)) are theoretically computed using the true rational curves $\mathbb{P}^1 \to V$. We have explained in § 6, following Witten, why we should have a formula of the type

$$Y_1(\omega)(\omega_1,\omega_2,\omega_3) = \int_V \omega_1 \wedge \omega_2 \wedge \omega_3 + \sum_{\{f\}} \exp\left(-\int_{\mathbb{P}^1} f^*\omega\right) \alpha(\{f\},\omega_1,\omega_2,\omega_3) \tag{5.45}$$

where the sum extends over all the components $\{f\}$ of the set of nonconstant holomorphic maps of $\mathbb{P}^1$ into V, the contribution $\alpha(\{f\},\omega_1,\omega_2,\omega_3)$ being obtained as an integral over the component $\{f\}$.

It is thus essentially necessary to understand the contribution of each component. One may reasonably hope, at least for certain Calabi–Yau threefolds, that for a general complex structure on V, all the generically embedded rational curves are rigid, that is, have no infinitesimal deformations. (For quintics of dimension 3, that is the content of a conjecture of Clemens.) Such curves $f : \mathbb{P}^1 \to V$ provide components $\{f\}$ of dimension 3 of the set of nonconstant holomorphic maps of $\mathbb{P}^1$ into V,

$$\{f\} = \{f \circ \phi; \phi \in \operatorname{Aut} \mathbb{P}^1\}.$$

Such a component admits $\mathbb{P}^3$ as a natural compactification. If the curve $f(\mathbb{P}^1)$ is infinitesimally rigid, we have $h^1\big((f \circ \phi)^* T_V\big) = \{0\}$. Thus $\{f\}$ appears as a reduced component of the right dimension of the locus of zeros of the section $s = \bar{\partial}\phi$ of the bundle W on the space M of maps ϕ of class $\mathcal{C}^\infty$ from $\mathbb{P}^1$ into V such that $\phi_*([\mathbb{P}]) = A$, whose fiber at ϕ is equal to $W_\phi = \mathcal{C}^\infty\big(\Omega^{0,1}_{\mathbb{P}^1} \otimes \phi^* T_V\big)$. In view of the form of the functional integral (2.53), the bosonic part of the action being equal to $2\|\bar{\partial}\phi\|^2$, Aspinwall and Morrison [70] essentially interpret the integral

$$(5.46) \qquad \int_{\phi_\alpha, \psi, \chi} \mathcal{O}_1(p_1)\mathcal{O}_2(p_2)\mathcal{O}_3(p_3) e^{-S'(\phi, \psi, \chi)} \, d\phi_\alpha \, d\psi \, d\chi,$$

(see (2.53)) as an integral

$$(5.47) \qquad \int_M s^* U \wedge \eta_1(p_1) \wedge \eta_2(p_2) \wedge \eta_3(p_3),$$

where U would be a Mathai–Quillen form [80] for the bundle W, the forms $\eta_i(p_i)$ being defined by $\eta_i(p_i) = \operatorname{ev}_{p_i}(\omega)$, where $\operatorname{ev}_{p_i} : M \to V$ sends ϕ to $\phi(p_i)$.

In this case, for a reduced component of the proper dimension $\{f\}$ of the locus of zeros of s, the contribution $\alpha(\{f\}, \omega_1, \omega_2, \omega_3)$ to the integral (5.47) is simply equal to $\int_{\{f\}} \eta_1(p_1) \wedge \eta_2(p_2) \wedge \eta_3(p_3)$.

On the other hand, by compactifying $\{f\}$ in $\mathbb{P}^3$ we can construct a diagram

$$(5.48) \qquad \begin{array}{ccccc} \Gamma_{p_i} & \xrightarrow{q_i} & \mathbb{P}^1 & \xrightarrow{f} & V \\ \downarrow{\scriptstyle q'_i} & & & & \\ \mathbb{P}^3 & & & & \end{array}$$

where the map q'_i is birational and q_i is a resolution of the singularities of the meromorphic map of $\mathbb{P}^3$ into $\mathbb{P}^1$ that assigns $\phi(p_i)$ to $\phi \in \operatorname{Aut} \mathbb{P}^1$.

It is then natural to define the class $\eta_i(p_i)$ on $\mathbb{P}^3$ by

$$\eta_i(p_i) = q'_{i*} \circ q_i^* \circ f^*\big([\omega_i]\big),$$

the term on the right being independent of the choice of the resolution.

When ω_i is of degree equal to 2, one can verify immediately the equality

$$\eta_i(p_i) = \int_{\mathbb{P}^1} f^*(\omega_i) c_1\big(\mathcal{O}_{\mathbb{P}^3}(1)\big).$$

We thus find that the contribution of such a component satisfies the equality

$$(5.49) \qquad \alpha\big(\{f\}, \omega_1, \omega_2, \omega_3\big) = \int_{\mathbb{P}^1} f^*\omega_1 \int_{\mathbb{P}^1} f^*\omega_2 \int_{\mathbb{P}^1} f^*\omega_3.$$

Unfortunately, and even if Clemens' conjecture is true, it is necessary to take account of the contribution of the components $\{f_k\}$ made up of the maps $g : \mathbb{P}^1 \to V$

of the form $g = f \circ \phi$, where f is of degree 1 on its image (which is an infinitesimally rigid curve) and ϕ is a map of degree k from $\mathbb{P}^1$ to $\mathbb{P}^1$.

Aspinwall and Morrison then propose, by analogy with the preceding, to calculate the contribution $\alpha(\{f_k\}, \omega_1, \omega_2, \omega_3)$ by the formula

$$(5.50) \qquad \alpha(\{f_k\}, \omega_1, \omega_2, \omega_3) = \int_{\{f_k\}} c_{2(k-1)}(E)\eta_1(p_1) \wedge \eta_2(p_2) \wedge \eta_3(p_3),$$

where E is the bundle (of rank $2(k-1)$) with fiber $E_\phi = H^1(\phi^*(T_V))$, which would be correct if $\{f_k\}$ were compact, since $H^1(\phi^*(T_V))$ can be identified with the cokernel of the map $ds: T_{M_\phi} \to W_\phi$.

To give a meaning to this expression, it is thus necessary to construct a compactification of $\{f_k\}$ to which the the bundle E can be extended and to extend the class of the forms $\eta_i(p_i)$.

Aspinwall and Morrison choose the simplest compactification, which is the projective space of dimension $2k+1$ of the sections of the bundle $\mathcal{O}_Q(k, 1)$ over the surface $Q = \mathbb{P}^1 \times \mathbb{P}^1$. Indeed, a generic element of $\mathbb{P}^{2k+1}$ parameterizes exactly one curve C on Q isomorphic to $\mathbb{P}^1$ by the first projection π_1 and of degree k over $\mathbb{P}^1$ by the second projection π_2, that is, a morphism $\pi_2 \circ \pi_{1|C}^{-1} : \mathbb{P}^1 \to \mathbb{P}^1$ of degree k. We can define the classes $\eta_i(p_i)$, as above, by the formula

$$(5.51) \qquad \eta_i(p_i) = q'_{i*} \circ q_i^* \circ f^*([\omega_i]),$$

where the map q_i is obtained by resolution of the singularities of the meromorphic map of $\mathbb{P}^{2k+1}$ into $\mathbb{P}^1$ that associates $\pi_2 \circ \pi_{1|C}^{-1}(p_i)$ with C:

$$(5.52) \qquad \begin{array}{ccccc} \Gamma_{p_i} & \xrightarrow{q_i} & \mathbb{P}^1 & \xrightarrow{f} & V \\ \downarrow q'_i & & & & \\ \mathbb{P}^{2k+1} & & & & \end{array}$$

We see immediately that

$$\eta_i(p_i) = \left(\int_{\mathbb{P}^1} f^* \omega_i\right) c_1\left(\mathcal{O}_{\mathbb{P}^{2k+1}}(1)\right).$$

On the open set $\mathbb{P}^{2k+1}$ made up of the morphisms of degree k from $\mathbb{P}^1$ into $\mathbb{P}^1$, we have a vector bundle E of rank $2(k-1)$ whose fiber at ϕ is $E_\phi = H^1(f \circ \phi^*(T_V))$. It can be extended to $\mathbb{P}^{2k+1}$ as follows:

Let $D \subset \mathbb{P}^{2k+1} \times Q$ be the universal divisor and pr_1 and pr_2 the projections of $\mathbb{P}^{2k+1} \times Q$ on its factors. We set

$$E = R^1 \mathrm{pr}_{1*}\left(\mathrm{pr}_2^*((f \circ \pi_2)^*(T_V))_{|D}\right).$$

Using the notation $F = (f \circ \pi_2)^* T_V$, we then have the following exact sequence over $\mathbb{P}^{2k+1} \times Q$:

$$(5.53) \qquad 0 \to \mathrm{pr}_2^*(F)(-1, -k, -1) \to \mathrm{pr}_2^*(F) \to \mathrm{pr}_2^*(F)_{|D} \to 0,$$

where $\mathcal{O}(-1, -k, -1) \cong \mathcal{I}_D$ is the bundle

$$\mathrm{pr}_1^*\left(\mathcal{O}_{\mathbb{P}^{2k+1}}(-1)\right) \otimes \mathrm{pr}_2^*\left(\mathcal{O}_Q(-k, -1)\right).$$

Since $H^1\left((f \circ \pi_2)^*(T_V)\right) = \{0\}$, we obtain immediately an isomorphism

$$(5.54) \qquad E = R^1 \mathrm{pr}_{1*}\left(\mathrm{pr}_2^*(F)_{|D}\right) \cong \mathcal{O}_{\mathbb{P}^{2k+1}}(-1) \otimes H^2(Q, F(-k, -1)).$$

From that we deduce the equality
$$c_{2(k-1)}(E) = c_1\big(\mathcal{O}_{\mathbb{P}^{2k+1}}(1)\big)^{2(k-1)}$$
and the formula
$$(5.55) \qquad \alpha\big(\{f_k\}, \omega_1, \omega_2, \omega_3\big) = \int_{\mathbb{P}^1} f^*\omega_1 \int_{\mathbb{P}^1} f^*\omega_2 \int_{\mathbb{P}^1} f^*\omega_3.$$

Formulas (5.45) and (5.55) now provide the following expression for $Y_1(\omega)(\omega_1, \omega_2, \omega_3)$, assuming that all the generically embedded curves are infinitesimally rigid

$$(5.56) \quad Y_1(\omega)(\omega_1, \omega_2, \omega_3)$$
$$= \int_V \omega_1 \wedge \omega_2 \wedge \omega_3 + \sum_{\substack{\mathbb{P}^1 \subset V \\ k \geq 1}} e^{-k \int_{\mathbb{P}^1} f^*\omega} \int_{\mathbb{P}^1} f^*\omega_1 \int_{\mathbb{P}^1} f^*\omega_2 \int_{\mathbb{P}^1} f^*\omega_3,$$

which, when the sum is extended over k and the number of the generically imbedded rational curves of class A is denoted $N(A)$, becomes

$$(5.57) \quad Y_1(\omega)(\omega_1, \omega_2, \omega_3)$$
$$= \int_V \omega_1 \wedge \omega_2 \wedge \omega_3 + \sum_A N(A) \frac{e^{-\int_A \omega}}{1 - e^{-\int_A \omega}} \int_A \omega_1 \int_A \omega_2 \int_A \omega_3,$$

that is, the formula used in § 3.

The reasoning described here is very incomplete, since it rests on the unjustified choice of the natural compactification of the space of maps of degree k from $\mathbb{P}^1$ into $\mathbb{P}^1$ given by the space $\mathbb{P}^{2k+1}$ in order to apply the excess formula (5.50).

Actually, the formula of Aspinwall and Morrison (5.57) was proved rigorously by Manin [79], using certain ideas of Kontsevich [51] and using Bott's fixed-point formula for smooth algebraic stacks, and more recently by Voisin [85], following a line of reasoning closer to that proposed by Aspinwall and Morrison. In these latter articles relation (5.57) is shown assuming the equality (which one can interpret as the definition of the left-hand side)

$$(5.58) \qquad Y_1(\omega)(\omega_i, \omega_j, \omega_k) = \frac{\partial^3 \Phi_\omega}{\partial t_i \partial t_j \partial t_k}(0),$$

in which $\{\omega_l\}$ is a basis of $H^2(X, \mathbb{C})$, and the t_l are the corresponding linear coordinates on $H^2(X, \mathbb{C})$. (The right-hand side is also equal to $\dfrac{\partial^3 \Phi_{\omega+\alpha}}{\partial t_i \partial t_j \partial t_k}(\alpha)$ for all $\alpha \in H^2(X, \mathbb{C})$.)

To do this we first show Proposition 5.11, which involves the calculation of the Gromov–Witten invariants. Let $j : \mathbb{P}^1 \to X$ be a rigid immersion (necessarily of degree 1 in its image), that is, such that the normal bundle $N_{\mathbb{P}^1}X$ is isomorphic to $\mathcal{O}_{\mathbb{P}^1}(-1) \oplus \mathcal{O}_{\mathbb{P}^1}(-1)$; let $A = j_*\big([\mathbb{P}^1]\big) \in H_2(X, \mathbb{Z})$ and k a strictly positive integer.

We consider a small generic deformation J of the almost-complex structure of X and a section ν near 0 of the bundle $\mathrm{pr}_1^*\Omega_{\mathbb{P}^1}^{0,1} \otimes \mathrm{pr}_2^* T_{X,J}^{1,0} \to \mathbb{P}^1 \times X$. We assume (J, ν) is generic.

The set $W_{kA,J,\nu}$ then possesses a component $W^0_{kA,J,\nu}$ made of maps $\phi : \mathbb{P}^1 \to X$ with image contained in a small neighborhood of $j(\mathbb{P}^1)$. This component "deforms" the family of holomorphic maps $\phi : \mathbb{P}^1 \to X$ of the form $j \circ f$ for $f : \mathbb{P}^1 \to \mathbb{P}^1$ of degree k.

The contribution of this family to the Gromov–Witten invariants will be (by definition) given by the image of the evaluation maps restricted to $W^0_{kA,J,\nu} \times (\mathbb{P}^1)^l$.

PROPOSITION 5.11. *Let x_1, x_2, x_3 be three distinct points of $\mathbb{P}^1$, and let* ev $: W^*_{kA,j,\nu} \to X^3$ *be the map defined by*

$$\mathrm{ev}\,(\phi) = \big(\phi(x_1), \phi(x_2), \phi(x_3)\big).$$

Then the closure of the image of ev *has homology class $A \otimes A \otimes A \in H_6(X^3, \mathbb{Z})$ (where we have used the natural orientation of $W^0_{kA,j,\nu}$ (see [81]))*.

Now assuming that all the rational curves generically embedded in X are embedded and rigid, we deduce easily from this proposition the following formula for the Gromov–Witten potential of X:
(5.59)
$$\Phi_\omega(\alpha) = \frac{1}{6}\int_X \alpha^3 + \sum_{0 \neq A \in H_2(X,\mathbb{Z})} N(A) \sum_{\substack{m \geq 3 \\ k \geq 1}} \frac{k^{m-3}}{m!}\Big(\int_A \alpha\Big)^m \exp\Big(-\int_{kA}\omega\Big),$$

where $N(A)$ is the number of rational curves generically embedded of class A. Differentiating this expression three times, we find
(5.60)
$$\frac{\partial^3 \Phi_\omega}{\partial t_i \partial t_j \partial t_k}(\alpha)$$

$$= \int_X \omega_i \omega_j \omega_k$$
$$+ \sum_{A \neq 0} N(A) \Big(\sum_{\substack{m \geq 3 \\ k \geq 1}} \frac{k^{m-3}}{(m-3)!}\Big(\int_A \alpha\Big)^{m-3} \exp\Big(-\int_{kA}\omega\Big)\Big)$$
$$\times \int_A \omega_i \int_A \omega_j \int_A \omega_k$$
$$= \int_X \omega_i \omega_j \omega_k + \sum_{A \neq 0} N(A) \Big(\sum_{k \geq 1} \exp\Big(\int_{kA} -\omega + \alpha\Big)\Big) \int_A \omega_i \int_A \omega_j \int_A \omega_k$$
$$= \int_X \omega_i \omega_j \omega_k + \sum_{A \neq 0} N(A) \frac{e^{\int_A -\omega + \alpha}}{1 - e^{\int_A -\omega + \alpha}} \int_A \omega_i \int_A \omega_j \int_A \omega_k,$$

which proves (5.57). □

CHAPTER 6

The Givental Construction

The purpose of the present chapter is on the one hand to provide some complementary material to the preceding chapter and on the other to explain the (partly rigorous, partly intuitive) ideas used by Givental [**91**] to reach the same conclusion as Candelas, de la Ossa, Green, and Parkes (see Chapter 3), namely that counting the rational curves on a quintic of $\mathbb{P}^4$ should make it possible to construct a function connected by simple transformations to the solutions of the Picard–Fuchs equation of the mirror family.

Floer cohomology, to which the first part of the chapter is devoted, is not actually used in the following part, which is devoted to the Givental construction, except to guarantee that certain formal expressions used there have a meaning.

We begin by presenting the Floer theory, which consists of using the existence of the Conley–Zehnder index to construct the Floer complex. We consider a certain functional defined on the covering of the loop space of a symplectic variety. The Conley–Zehnder index is a substitute for the Morse index; it gives the dimension of the space of trajectories of the flow of the symplectic gradient of this functional joining two critical points.

The term of order 1 in this flow is given by the Cauchy–Riemann equation for a generic pseudocomplex structure. The cohomology of the Floer complex is essentially comparable to the cohomology of the variety in question having coefficients in a certain ring of formal series, and it can be endowed with a natural product, which has recently been identified with the quantum product.

We then explain the elements of the equivariant cohomology, and in particular of the S^1-equivariant cohomology of a symplectic variety endowed with a Hamiltonian action, which are necessary to understand the Givental calculation. Following [**91**], we explain the construction of a $\mathcal{D}$-module structure on $H^*_{S^1}(M)$ under certain assumptions, and also the calculation of certain equivariant Euler classes on the spaces of Laurent polynomials $P(z)$ with coefficients in $\mathbb{C}^5$ endowed with the action of S^1 given by the loop rotation $P(z) \mapsto P(\lambda z)$ with $\lambda \in S^1$, which, by passage to the limit, provide formal solutions of the Picard–Fuchs equation mentioned above.

1. Floer Cohomology

This theory was created by Floer [**89**], [**90**] to obtain Morse inequalities for the number of fixed points of an exact diffeomorphism of a symplectic manifold (that is, a diffeomorphism obtained by integrating a time-dependent field $X(t)$ that is globally Hamiltonian for all times t) or for the number of periodic orbits of a periodic Hamiltonian flow (Arnold's conjecture). Let (V, ω) be a compact symplectic manifold and H a $\mathcal{C}^\infty$-function on $\mathbb{R} \times V$ such that $H(t, v) = H(t+1, v)$.

Through the formula
$$\text{int}_{X_H(t)}(\omega) = dH_t$$
this function determines a time-dependent Hamiltonian field X_H on V, and hence a flow
$$\psi_t : V \to V.$$
The relation $\psi_1(v) = v$ holds if and only if $\psi_t(v)$ is a periodic orbit of X_H.

Let LV be the space of loops homotopic to a constant and $\widetilde{LV}$ the covering of LV corresponding to the kernel of the composite map

(6.1) $$\pi_1(LV) \to \pi_2(V) \to H_2(V) \xrightarrow{(c_1,\omega)} \mathbb{R}^2$$

where $c_1, \omega : H_2(V, \mathbb{Z}) \to \mathbb{R}$ are defined by integrating $c_1(V), \omega \in H^2(V)$.

The space $\widetilde{LV}$ can be identified with the space of maps $\phi : D^2 \to V$ modulo the relations
$$\phi \equiv \psi \text{ if } \phi_{|\partial D^2} = \psi_{|\partial D^2}, \quad \int_{D^2} \phi^* c_1 = \int_{D^2} \psi^* c_1, \quad \int_{D^2} \phi^* \omega = \int_{D^2} \psi^* \omega,$$
where D^2 is the unit disk in $\mathbb{R}^2$. The form ω and the function H make it possible to define the following action on $\widetilde{LV}$:

(6.2) $$\mathcal{A}_H(\phi) = \int_{D^2} \phi^*(\omega) + \int_{S^1} H(t, \phi(t)) \, dt,$$

where $S^1 = \partial D^2$ is identified with $\mathbb{R}/\mathbb{Z}$.

The critical points of $\mathcal{A}_H$ are defined by the condition

(6.3) $$\forall \xi \in \mathcal{C}^\infty(\phi_{|S^1}^*(TV)), \quad \int_{S^1} \text{int}_\xi \omega + \int_{S^1} d_V H(t, \phi(t))(\xi) \, dt = 0,$$

which is clearly equivalent to
$$\omega(\phi'(t), \bullet) = d_V H(t, \phi(t))(\bullet)$$
and also to the fact that $\phi_{|S^1}$ is a periodic orbit of X_H.

Now let J be a pseudocomplex structure on V compatible with ω. We associate with (J, ω) the metric $g(u, v) = \omega(u, Jv)$ on TV and hence the metric L^2 on T_{LV} (or $T_{\widetilde{LV}}$) defined as follows. For $\phi \in LV$ the tangent space to LV at ϕ is the space of $\mathcal{C}^\infty$-sections of the bundle $\phi^* T_V$. Let $\rho, \xi \in \mathcal{C}^\infty(\phi^*(TV))$. We set
$$\langle \rho, \xi \rangle_\phi = \int_{S^1} g_{\phi(t)}(\rho(t), \xi(t)) \, dt.$$

Because of the relation
$$d\mathcal{A}_H(\xi) = \int_{S^1} \omega(\xi(t), \phi'(t)) + \omega(X_{H_t}(\phi(t)), \xi(t)) \, dt,$$
we see that the gradient of $\mathcal{A}_H$ for this metric is the field
$$\phi \mapsto -J\phi'(t) + JX_H(t) \in \mathcal{C}^\infty(\phi^*(TV)).$$

Even if one assumes that the critical points of $\mathcal{A}_H$ are nondegenerate, it is not possible to do Morse theory with this functional, since its Hessian at a critical point has an infinite number of positive and negative eigenvalues.

By replacing the Morse index with the Conley–Zehnder index, Floer nevertheless constructs an analog of the Thom–Smale complex (assuming that (V, ω) is monotone).

1.1. The Conley–Zehnder index. Let $\mathcal{H} \subset LV$ be the set of periodic orbits of X_H, let $\phi \in \mathcal{H}$, and let $\tilde{\phi} : D^2 \to V$ be a $\mathcal{C}^\infty$-map that extends ϕ. The symplectic bundle $\tilde{\phi}^* T_V$ can be trivialized over D^2. On the other hand, for $t \in S^1$ the differential
$$\psi_{T_*} : (\phi^* T_V)_0 \to (\phi^* T_V)_t$$
of the flow ψ_t provides a symplectic isomorphism, since X_H is Hamiltonian. By the trivialization induced by $\phi^* T_V$ the differential ψ_{t_*} provides a map
$$k_t : S^1 \longrightarrow \mathrm{Sp}_n \qquad (\dim V = 2n)$$
whose homotopy class is a measured by an integer $\mu(\tilde{\phi}, H)$ called the *Conley–Zehnder index* of $(\tilde{\phi}, H)$.

There exists a natural action
$$(u, \tilde{\phi}) \mapsto u \# \tilde{\phi}$$
of the group $\mathrm{Im}\bigl(\pi_2(v) \to H_2(V)\bigr)$ on $\widetilde{LV}$, which can be described as follows. If the class u of a map $\tilde{u} : S^2 \to V$ is given and $\tilde{\phi} \in \widetilde{LV}$ is represented by a map $\tilde{\phi} : D^2 \to V$ denoted in the same way, we may assume that for a base point x_0 in S^2 we have $\tilde{u}(x_0) = \tilde{\phi}(0)$. By identifying $S^2 - \{x_0\}$ with the interior of $D^2_{1/2}$ and mapping $D^2 - D^2_{1/2}$ onto D^2 by a map β of the form
$$z \mapsto \rho(|z|)z, \text{ with } \rho \text{ monotone, } \rho(1) = 1, \text{ and } \rho\Bigl(\frac{1}{2}\Bigr) = 0,$$
we define $u \# \tilde{\phi}$ as the class of the map of D^2 into V that is equal to $\tilde{u}$ on $D^2_{1/2}$ and to $\tilde{\phi} \circ \beta$ on $D^2 - D^2_{1/2}$.

The Conley–Zehnder index behaves as follows with respect to this action (which of course leaves the critical points of $\mathcal{A}_H$ invariant):

(6.4) $$\mu(u \# \tilde{\phi}, H) = \mu(\tilde{\phi}, H) + 2 \int_{S^2} u^* c_1(V).$$

The following result is due to Salamon and Zehnder.

THEOREM 6.1. *Let $\tilde{\phi}$ and $\tilde{\psi}$ be nondegenerate critical points of $\mathcal{A}_H$. Then for a generic pseudocomplex structure J the space $\mathcal{M}(\tilde{\phi}, \tilde{\psi})$ of solutions of the equation*

(6.5) $$\frac{\partial u}{\partial s}(s, t) = -J \frac{\partial u}{\partial t}(s, t) + J X_{H_t}\bigl(u(s, t)\bigr)$$

for $(s, t) \in \mathbb{R} \times S^1$ (that is, the trajectories of the gradient of $\mathcal{A}_H$) satisfying

(6.6) $$\lim_{s \to -\infty} u(s, t) = \phi, \quad \lim_{s \to +\infty} u(s, t) = \psi, \quad \tilde{\psi} = u \# \tilde{\phi} \in \widetilde{LV}$$

is of dimension $\mu(\tilde{\psi}, H) - \mu(\tilde{\phi}, H)$.

Here we have used the same notation $\#$ for the following operation. If
- $\tilde{\phi}$ is the equivalence class of a map $\tilde{\phi} : D^2 \to V$ satisfying $\tilde{\phi}_{|\partial D^2} = \phi$,
- u is a map of $\overline{\mathbb{R}} \times S^1$ into V satisfying $u_{|-\infty} = \phi$,
- $u_{|+\infty} = \psi$,

then $u\#\tilde{\phi}$ denotes the class in $\widetilde{LV}$ of the map of D^2 into V that equals $\tilde{\phi}(2z)$ on $D^2_{1/2}$ and u on $D^2 - D^2_{1/2}{}^0 \cong \overline{\mathbb{R}} \times S^1$.

In particular, for $\mu(\tilde{\psi}, H) - \mu(\tilde{\phi}, H) = 1$, we find isolated trajectories (modulo translations in time s). As in Morse theory, these trajectories can be used to construct the differential of the Floer complex. However, one does encounter the following difficulty: $\mathcal{A}_H$ may have an infinite number of critical points of index k in $\widetilde{LV}$. (This happens, for example, if $c_1 = 0$.)

The introduction of the Novikov ring makes it possible to overcome this difficulty.

1.2. The Novikov ring. Let Γ be a commutative group and $\phi : \Gamma \to \mathbb{R}$ a homomorphism. Given a ring R of coefficients $(\mathbb{Z}, \mathbb{Q}, \mathbb{R}, \dots)$, we define $\Lambda(\Gamma, \phi, R)$ as the set of functions $\rho : \Gamma \to R$ such that for every real number $c > 0$ the set

$$\{A \in \Gamma;\ \rho(A) \neq 0,\ \phi(A) < c\} \text{ is finite.}$$

The set $\Lambda(\Gamma, \phi, R)$ is a ring under addition of functions and the multiplication defined by the following formula, where the sum on the right is finite:

$$\rho \cdot \rho'(A) = \sum_B \rho(B)\rho'(A - B).$$

If Γ is a free $\mathbb{Z}$-module of finite type, with basis $(e_1, \dots, e_m)$, we can identify $\Lambda(\Gamma, \phi, R)$ with the set of series

$$\sum_{I \in \mathbb{Z}^m} \lambda_I t^I, \quad \lambda_I \in R,$$

in the formal variables $t_1, \dots, t_m$ satisfying the condition that for every real $c > 0$ the set

$$\left\{I = (i_1, \dots, i_m);\ \lambda_I \neq 0,\ \sum_k i_k \phi(e_k) < c\right\}$$

is finite.

The Novikov ring Λ_ω that will be used is constructed as above on the group

$$\Gamma = \operatorname{Ker} c_1 / \operatorname{Ker} c_1 \cap \operatorname{Ker} \omega,$$

endowed with the homomorphism ω. (By abuse of notation we use c_1 and ω for maps with values in $\mathbb{R}$ defined on $\pi_2(V)$ as the compositions of the natural arrow $\pi_2(V) \to H_2(V)$ and the linear forms on $H_2(V)$ corresponding to the classes ω and c_1 of $H^2(V)$.)

We note that if $c_1 = 0$, $\pi_2(V) = H_2(V)$, and $\omega : H_2(V, \mathbb{Z}) \to \mathbb{R}$ is one-to-one, then $\Gamma = H_2(V, \mathbb{Z})$.

1.3. The Floer complex. Since we have need of a compactness result for the image of the manifolds $\mathcal{M}(\tilde{\phi}, \tilde{\psi})$ under the evaluation map when $\mu(\tilde{\psi}, H) - \mu(\tilde{\phi}, H) \leq 2$, we assume henceforth that V is monotone (see [**89**]) or weakly monotone (see [**94**]), which avoids the bubble phenomenon for small indices and makes it possible to show the following.

THEOREM 6.2. *For $c > 0$ and $k \leq 2$, let*

$$\mathcal{M}_c^k \subset \bigcup_{\substack{\tilde{\phi}, \tilde{\psi} \\ \mu(\tilde{\psi}, H) \leq \mu(\tilde{\phi}, H) + k}} \mathcal{M}(\tilde{\phi}, \tilde{\psi})$$

be defined by the condition

$$E(u) := \frac{1}{2} \int_{-\infty}^{+\infty} \int_{S^1} \left|\frac{\partial u}{\partial s}\right|^2 + \left|\frac{\partial u}{\partial t} - X_{H_t}(u(s,t))\right|^2 ds\, dt \leq c.$$

Then the image in V of the map

$$\mathrm{ev} : \overline{\mathbb{R}} \times S^1 \times \mathcal{M}_c^k \longrightarrow V,$$
$$\mathrm{ev}(s,t,u) = u(s,t)$$

is compact.

(We have set $\overline{\mathbb{R}} = \mathbb{R} \cup \{-\infty, +\infty\}$ and extended u by passing to the limit.)

The Floer complex is then constructed as follows:

Let $\widetilde{\mathcal{H}}$ be the set of critical points of $\mathcal{A}_H$, and let $\widetilde{\mathcal{H}}_k$ be the set of critical points of index k. We define C^k as the set of functions $\xi : \widetilde{\mathcal{H}}_k \to R$ such that for every real $c > 0$, the set

$$\{\tilde{\phi} \in \widetilde{\mathcal{H}}_k;\ \xi(\tilde{\phi}) \neq 0,\ \mathcal{A}_h(\tilde{\phi}) \leq c\}$$

is finite. (We shall use the notation $\xi = \sum_{\tilde{\phi}} \xi(\tilde{\phi})\langle\tilde{\phi}\rangle$ below.) We now have

LEMMA 6.3. *The formula*

(6.7) $$\rho\xi(\tilde{\phi}) = \sum_{A \in \Gamma} \rho(A)\xi\big((-A)\#\tilde{\phi}\big),$$

where $\xi \in C^k$ and $\tilde{\phi} \in \mathcal{H}_k$, defines a natural action of Λ_ω on C^k.

Indeed, since $A \in \mathrm{Ker}\,c_1$, we have $(-A)\#\tilde{\phi} \in \widetilde{\mathcal{H}}_k$ by (6.4). On the other hand, the sum is finite since $\mathcal{A}_H\big((-A)\#\tilde{\phi}\big) = -\omega(A) + \mathcal{A}_H(\tilde{\phi})$. In order for $\rho(A)\xi\big((-A)\#\tilde{\phi}\big)$ to be nonzero, it is necessary that $\rho(A) \neq 0$ and $\xi\big((-A)\#\tilde{\phi}\big) \neq 0$. If we had an infinite number of nonzero terms, we would have a sequence A_i with

$$\lim_{i \to \infty} \omega(A_i) = \infty \text{ and } \xi\big((-A_i)\#\tilde{\phi}\big) \neq 0,$$

which is absurd, since $\lim_{i \to \infty} \mathcal{A}_H\big((-A_i)\#\tilde{\phi}\big) = -\infty$. This shows that $\rho \cdot \xi$ is indeed a function on $\widetilde{\mathcal{H}}_k$ with values in R. It can be shown similarly that we have $\rho \cdot \xi \in C^k$. □

We note now that C^k is actually a module of finite rank over Λ_ω (see [**94**]). Indeed, let N be the minimal Chern number of (V,ω), defined by

$$N\mathbb{Z} = \mathrm{Im}\,(c_1 : \pi_2(V) \to \mathbb{Z}).$$

Let $\mathcal{H}_k$ be the set of periodic orbits ϕ such that $\mu(\tilde{\phi}) \equiv k \mod 2N$, for a lifting $\tilde{\phi}$ of ϕ in $\widetilde{\mathcal{H}}$.

For $\phi \in \mathcal{H}_k$, we choose a lifting $\tilde{\phi}$ in $\widetilde{\mathcal{H}}_k$. We see immediately that the corresponding elements $\langle\tilde{\phi}\rangle \in C^k$ which is the function on $\widetilde{\mathcal{H}}_k$ taking the value 1 on $\tilde{\varphi}$ and 0 elsewhere, form a basis of C^k over Λ_ω. □

For $\tilde{\phi}, \tilde{\psi} \in \widetilde{\mathcal{H}}$, let $u \in \mathcal{M}(\tilde{\phi}, \tilde{\psi})$. We then have

(6.8) $$E(u) = \mathcal{A}_H(\tilde{\psi}) - \mathcal{A}_H(\tilde{\phi}).$$

By Theorems 6.1 and 6.2 and formula (6.8), there exists a finite number of trajectories in $\mathcal{M}(\tilde{\phi}, \tilde{\psi})$ modulo translations of time for $\tilde{\phi} \in \widetilde{\mathcal{H}}_k$ and $\tilde{\psi} \in \widetilde{\mathcal{H}}_{k+1}$. We can also count them with signs (see [**89**], [**94**]), which provides a number $n(\tilde{\phi}, \tilde{\psi}) \in \mathbb{Z}$. We then have the following result.

LEMMA 6.4. *The formula*
$$\partial \xi(\tilde{\psi}) = \sum_{\tilde{\phi} \in \tilde{\mathcal{H}}_k} n(\tilde{\phi}, \tilde{\psi}) \xi(\tilde{\phi}) \tag{6.9}$$

where $\xi \in C^k$ and $\tilde{\psi} \in \tilde{\mathcal{H}}_{k+1}$, defines a map
$$\partial : C^k \to C^{k+1}. \tag{6.10}$$

It is this map that will be the differential of the Floer complex.

We remark first of all that the sum in (6.9) is finite. Indeed, when u belongs to $\mathcal{M}(\tilde{\phi}, \tilde{\psi})$, we have
$$E(u) = \mathcal{A}_H(\tilde{\psi}) - \mathcal{A}_H(\tilde{\phi}) \geq 0,$$
and thus $\mathcal{A}_H(\tilde{\phi}) \leq \mathcal{A}_H(\tilde{\psi})$ when $\mathcal{M}(\tilde{\phi}, \tilde{\psi})$ is nonempty, and there exists a finite number of $\tilde{\phi} \in \tilde{\mathcal{H}}_k$ satisfying $\mathcal{A}_H(\tilde{\phi}) \leq \mathcal{A}_H(\tilde{\psi})$ and $\xi(\tilde{\phi}) \neq 0$.

One can verify likewise that $\partial \xi$ is indeed in C^{k+1}. If $\mathcal{A}_H(\tilde{\psi}) \leq c$, the condition $\partial \xi(\tilde{\psi}) \neq 0$ implies $\xi(\tilde{\phi}) \neq 0$ for an element $\tilde{\phi}$ of $\tilde{\mathcal{H}}_k$ such that $\mathcal{M}(\tilde{\phi}, \tilde{\psi})$ is nonempty.

We then have $\mathcal{A}_H(\tilde{\phi}) \leq c$, and there exists a finite number of such $\tilde{\phi}$. On the other hand, by Theorem 6.2 and formula (6.8), for such a $\tilde{\phi}$ the set
$$\bigcup_{\mathcal{A}_H(\tilde{\psi}) \leq c} \mathcal{M}(\tilde{\phi}, \tilde{\psi})$$
is finite modulo time translations, so that the set
$$\{\tilde{\psi}; \mathcal{A}_H(\tilde{\psi}) \leq c, \partial \xi(\tilde{\psi}) \neq 0\}$$
is indeed finite. □

We now have the following result:

THEOREM 6.5. *The differential ∂ satisfies $\partial^2 = 0$, and hence makes it possible to define the Floer cohomology groups*
$$FH^k(V, \omega) = \operatorname{Ker}(\partial : C^k \to C^{k+1}) / \operatorname{Im}(\partial : C^{k-1} \to C^k). \tag{6.11}$$

We note finally that ∂ commutes in an obvious manner with the action of Λ_ω, so that $FH^k(V, \omega)$ is a Λ_ω-module.

REMARK 6.6. The action of the covering group on $\widetilde{LV}$ provides an isomorphism
$$A\# : \tilde{\mathcal{H}}_k \longrightarrow \tilde{\mathcal{H}}_{k+2c_1(A)}$$
for $A \in \pi_2(V)$. On the other hand $A\#$ clearly preserves the trajectories, so that for each k we have an isomorphism (indeed canonical):
$$FH^k(V, \omega) \cong FH^{k+2N}(V, \omega),$$
where N is, as before, the the minimal Chern number of (V, ω).

It can be shown (see [89], [94]) that $FH^k(V,\omega)$ is independent of the generic choice of (H,J) (which justifies the notation *a posteriori*) in the sense that for (H_1, J_1), (H_2, J_2) satisfying the necessary conditions for the construction of the Floer complex we have a canonical isomorphism:

$$FH^k(V,\omega)_{(H_1,J_1)} \cong FH^k(V,\omega)_{(H_2,J_2)}.$$

2. The comparison theorem

Floer (in the monotone case and Hofer-Salamon (in the weakly monotone case) have shown the following result

THEOREM 6.7. *There exists a canonical isomorphism*

(6.12) $$FH^k(V,\omega) \cong \bigoplus_{l \equiv k \bmod 2N} H^{n+l}(V, \Lambda_\omega).$$

The proof is based on the fact that $FH^k(V,\omega)$ is independent of (H,J) and on the study of the case in which H is independent of time. It can then be shown that if H is a function that is sufficiently small on V, the only periodic orbits of X_H are the constant loops whose image is a critical point of H.

The second important point is the fact that the solutions $u(s,t)$ of the equation (6.5), $s \in \mathbb{R}$, $t \in S^1$, isolated modulo time-translation, with

$$\lim_{s \to -\infty} u(s,t) = c_x, \quad \lim_{s \to +\infty} u(s,t) = c_y,$$

where c_x and c_y are constant loops whose images are the critical points x, y of H, are independent of t, and hence can be identified with the trajectories of the gradient JX_H of H between the critical points x and y.

To conclude, it is necessary to compare the Conley–Zehnder index of $\tilde{\phi}_x$, where x is a critical point of H and $\tilde{\phi}_x : D^2 \to V$ is the constant map $\tilde{\phi}_x(z) = x$, with the Morse index of H at the point x.

We have the following relation, shown by Salamon and Zehnder:

(6.13) $$\mu(\tilde{\phi}_x, H) = \text{ind}_H(x) - n \quad (\dim V = 2n).$$

It follows from the preceding that if (C^*, ∂) and (M^*, ∂) are the Floer complexe of $\mathcal{A}_H$ and the Thom–Smale complex of H respectively, we have a natural isomorphism of complexes

(6.14) $$C^k \cong \bigoplus_{j \equiv k \bmod 2N} M^{j+n} \otimes \Lambda_\omega.$$

This provides the isomorphism (6.12).

REMARK 6.8. Givental and Kim [93] propose obtaining this isomorphism by considering the case in which $H = 0$. In this case, the set of critical points of $\mathcal{A}_H$ is made up of an infinite number of copies of the manifold V, the constant loops being the periodic orbits of $X_H = 0$. They then propose extending the Floer theory by analogy with the Bott extension of Morse theory by taking as the Floer homology cycles the cycles C_A, where $A \subset V$ represents an element of $H_*(V, \mathbb{Z})$, defined by

$$C_A = \{u(t), t \in S^1; u(t) = u(s_0, t)$$

where $u(s,t)$ satisfies (6.5) with $H = 0$ and

$$\lim_{s \to -\infty} u(s,t) \text{ is a constant loop with image in } A\}.$$

They then claim that the zero differential on the space generated by these cycles provides the correct definition of Floer homology. This approach was rigorously justified by Piunikhin, Salamon, and Schwarz [96]. An interesting point in this approach is the fact that the trajectories $u(s,t)$ used above are simply pseudo-holomorphic disks in V intersecting the cycles A in 0.

3. Quantum cohomology and Floer cohomology

The purpose of the present section is partly to complement Chapter 5 by showing that the quantum product can be defined formally provided one introduces the adequate ring of coefficients and partly to show that this product can be interpreted as a natural product on the Floer cohomology.

It will not be used in the rest of the chapter.

3.1. The Novikov ring and the quantum product. Here we follow [81].

In Section 5 we defined a product "$\bullet_\omega$" on the cohomology of a weakly monotone variety, under the hypothesis that the series considered converge, by the formula

$$(6.15) \qquad \langle \alpha \bullet_\omega \beta, \gamma \rangle = \sum_{A \in H_2(V, \mathbb{Z})} e^{-\omega(A)} \Phi_A(\alpha, \beta, \gamma).$$

We note that by definition $\Phi_A(\alpha, \beta, \gamma)$ can be nonzero only if

$$A \in \mathrm{Im}\,(\pi_2 \to H_2) \text{ and } \deg \alpha + \deg \beta + \deg \gamma = 2(n + c_1(A)).$$

On the other hand, by compactness (see [76]), we have the following result. *For all real $c > 0$ there exists a finite number of classes $A \in H_2(V, \mathbb{Z})$ such that $\omega(A) < c$ and $\Phi_A(\alpha, \beta, \gamma) \neq 0$ for at least one triplet (α, β, γ) of elements of V.*

Let Λ'_ω be the Novikov ring constructed as in § 1.2 on the group

$$\Gamma' = \pi_2(V)/\operatorname{Ker} c_1 \cap \operatorname{Ker} \omega \text{ with } R = \mathbb{Q}.$$

The ring Λ'_ω can be identified with the ring of formal series $\sum_{A \in \Gamma'} \lambda_A e^A$ with $\lambda_A \in \mathbb{Q}$ satisfying the following condition: *for every $c > 0$ there exists a finite number of classes $A \in \Gamma'$ such that $\omega(A) < c$ and $\lambda_A \neq 0$*, endowed with the multiplication

$$(\lambda \cdot \lambda')_A = \sum_{B+C=A} \lambda_B \lambda'_C \quad \text{(the sum is finite.)}$$

The ring Λ'_ω is graded by $\deg(e^A) = 2c_1(A)$. We consider $H^*(V, \mathbb{Q}) \otimes \Lambda'_\omega$ endowed with the grading $\deg(\alpha \otimes e^A) = \deg \alpha + 2c_1(A)$; we note that we have

$$(6.16) \qquad (H^*(V, \mathbb{Q}) \otimes \Lambda'_\omega)^k \cong \bigoplus_{l \equiv k \bmod 2N} H^l(V, \mathbb{Q}) \otimes \Lambda_\omega \cong FH^{k-n}(V, \omega),$$

where the last isomorphism is that of (6.12). We now have the following formal version of the quantum product which was defined in the preceding chapter only assuming convergence:

PROPOSITION 6.9. *The formula*

$$(6.17) \qquad \langle \alpha \bullet \beta, \gamma \rangle = \sum_{A \in \mathrm{Im}\,(\pi_2(V) \to H_2(V))} e^{\bar{A}} \Phi_A(\alpha, \beta, \gamma),$$

where $\bar{A}$ is the projection of A on Γ', defines a graded associative product over $H^*(V, \mathbb{Q}) \otimes \Lambda'_\omega$ if we make the following definition for $\alpha, \beta, \gamma \in H^*(V, \mathbb{Q})$:

$$\Phi_A\big(\alpha \otimes e^{A_1}, \beta \otimes e^{B_1}, \gamma \otimes e^{C_1}\big) = e^{A_1+B_1+C_1}\Phi_A(\alpha, \beta, \gamma).$$
$$\langle \alpha \otimes e^A, \beta \otimes e^B \rangle = e^{A+B}\langle \alpha, \beta \rangle.$$

PROOF. Formula (6.17) is equivalent to

(6.18) $$\alpha \otimes e^{A_1} \bullet \beta \otimes e^{B_1} = \sum_{A,\sigma,\tau} g^{\sigma\tau} \Phi_A(\alpha, \beta, e_\sigma) \otimes e_\tau e^{\bar{A}+A_1+B_1}$$

where e_σ is a basis of $H^*(V, \mathbb{Q})$ and $g^{\sigma\tau}$ is the inverse of the intersection matrix. By compactness the right-hand side contains only a finite number of nonzero coefficients

$$\sum_{A,\sigma} g^{\sigma\tau}\Phi_A(\alpha, \beta, e_\sigma)$$

for fixed τ, and $\omega(\bar{A} + A_1 + B_1) < c$, for every $c \in \mathbb{R}$, so that this product really is in $H^*(V, \mathbb{Q}) \otimes \Lambda'_\omega$. Finally, assuming that α, β, e_σ are homogeneous, we have

$$\Phi_A(\alpha, \beta, e_\sigma) = 0 \text{ if } \deg \alpha + \deg \beta + \deg e_\sigma \neq 2\big(n + c_1(A)\big)$$

and thus we have

$$\deg\big(\alpha \otimes e^{A_1} \bullet \beta \otimes e^{B_1}\big) = \deg \alpha + \deg \beta + 2\big(c_1(A_1) + c_1(B_1)\big)$$

since for $\Phi_A(\alpha, \beta, e_\sigma) \neq 0$ and $g^{\sigma\tau} \neq 0$, we have:

$$\begin{cases} \deg e_\tau = 2n - \deg \sigma, \\ \deg e_\tau \otimes e^{A+A_1+B_1} = 2\big(c_1(A) + c_1(A_1) + c_1(B_1)\big) + \deg e_\tau, \\ \deg e_\sigma = 2\big(n + c_1(A)\big) - \deg \alpha - \deg \beta. \end{cases}$$

The proof of associativity is carried out as in § 5. □

3.2. The product on the Floer cohomology. We consider three functions H_i, $i = 1, 2, 3$, on $S^1 \times V$ and an almost-complex structure J on V, making it possible to construct the Floer complexes (C_i^k, ∂), as in § 1.3. For each i we choose a function $H_i'(s, t, v)$ on $\mathbb{R} \times S^1 \times V$ such that

$$H_i'(s, t, v) = \begin{cases} H_i(t, v) & \text{for } |s| \geq 1, \\ H_i'(s, t, v) = 0 & \text{for } |s| \leq \frac{1}{2}. \end{cases}$$

Let Σ be the sphere S^2 with three disjoint disks D_i removed. We choose conformal parameterizations of neighborhoods U_i of ∂D_i in Σ:

(6.19) $$\begin{cases} \eta_i :]-\infty, 0[\times S^1 \cong U_i, \text{ for } i = 1, 2, \\ \eta_3 :]0, +\infty[\times S^1 \cong U_3, \end{cases}$$

with $\lim_{s \to -\infty} \eta_i(s, t) \in \partial D_i$ for $i = 1, 2$ and $\lim_{s \to +\infty} \eta_3(s, t) \in \partial D_3$.

Let $\tilde{\phi}_i$, $i = 1, 2, 3$, be critical points of $\mathcal{A}_{H_i}$ in $\widetilde{LV}$ and ϕ_i the corresponding periodic orbits of X_{H_i}. We consider the space $\mathcal{M}(\tilde{\phi}_1, \tilde{\phi}_2, \tilde{\phi}_3)$ of the solutions $u : \Sigma \to V$ of the equation

(6.20) $$\begin{cases} \dfrac{\partial u_i}{\partial s}(s, t) = -J\dfrac{\partial u_i}{\partial t}(s, t) + JX_{H_i'(s,t)}\big(u(s, t)\big), \\ u_{|\Sigma - \cap U_i} \text{ pseudoholomorphic,} \end{cases}$$

where $u_i = u \circ \eta_i$ and thus $s \leq 0$, $t \in S^1$ for $i = 1, 2$ and $s \geq 0$, $t \in S^1$ for $i = 3$, satisfying

(6.21) $$\begin{cases} \lim_{s \to -\infty} u_i(s,t) = \phi_i \text{ for } i = 1, 2, \quad \lim_{s \to +\infty} u_3(s,t) = \phi_3, \\ \tilde{\phi}_3 = u \#(\tilde{\phi}_1 \sqcup \tilde{\phi}_2) \in \widetilde{LV}. \end{cases}$$

Since $H'_i(s, t, v)$ is zero for $|s| \leq \frac{1}{2}$, the function u_i is pseudoholomorphic for $|s| \leq \frac{1}{2}$. On the other hand, since $H'_i(s, t, v) = H_i(t, v)$ for $|s| \geq 1$, we have

$$\begin{cases} \lim_{s \to -\infty} u_i(s,t) \text{ is a periodic orbit of } X_{H_i} \text{ for } i = 1, 2, \\ \lim_{s \to +\infty} u_3(s,t) \text{ is a periodic orbit of } X_{H_3}, \end{cases}$$

so that these equations are compatible.

It can be shown that for a generic choice of H'_i and J the space $\mathcal{M}(\tilde{\phi}_1, \tilde{\phi}_2, \tilde{\phi}_3)$ is of dimension $\mu(\tilde{\phi}_3, H_3) - \mu(\tilde{\phi}_1, H_1) - \mu(\tilde{\phi}_2, H_2) - n$. When

$$\mu(\tilde{\phi}_3, H_3) - \mu(\tilde{\phi}_1, H_1) - \mu(\tilde{\phi}_2, H_2) - n = 0,$$

we can count these solutions with an adequate sign, making it possible to construct a series of maps

(6.22) $$\begin{cases} \nu_{k,l} : C^k_{H_1} \times C^l_{H_2} \longrightarrow C^{k+l+n}_{H_3}, \\ \nu_{k,l}(\langle \tilde{\phi} \rangle, \langle \tilde{\psi} \rangle) = \sum_{\tilde{\chi}} n(\tilde{\phi}, \tilde{\psi}, \tilde{\chi}) \langle \tilde{\chi} \rangle \end{cases}$$

where it can be shown by a compactness argument that the term on the right really is an element of $C^{k+l+n}_{H_3}$.

It can be shown that the maps $\nu_{k,l}$ commute with the differentials and thus make it possible to construct a product $\eta = \bigoplus \eta_{k,l}$

(6.23) $$\eta_{k,l} : FH^k(V, \omega) \otimes FH^l(V, \omega) \longrightarrow FH^{k+l+n}(V, \omega).$$

By the isomorphism (6.16), η provides a graded product on $H^*(V, \mathbb{Q}) \otimes \Lambda'_\omega$

(6.24)
$$\eta : \left(H^*(V, \mathbb{Q}) \otimes \Lambda'_\omega\right)^{k+n} \otimes \left(H^*(V, \mathbb{Q}) \otimes \Lambda'_\omega\right)^{l+n} \longrightarrow \left(H^*(V, \mathbb{Q}) \otimes \Lambda'_\omega\right)^{k+l+2n}.$$

The following comparison theorem was established by Piunikhin, Salamon, and Schwarz in [**96**].

THEOREM 6.10. *The product η coincides with the quantum product (6.17).*

REMARK 6.11. If we assume the approach of Givental and Kim, which amounts to setting $H_i = 0$, the identity of the two products is heuristically clear, since the cycles that they use to construct the Floer homology are of the form

$$C_{A_\alpha} = \{u(t)_\alpha, t \in S^1; \exists \tilde{u} : D^2 \to V \text{ pseudoholomorphic}$$
$$\text{with } u(t) = \tilde{u}_{|\partial D^2}, u(0) \in A \text{ and } u(t)_\alpha = \alpha \# \tilde{u} \text{ in } \widetilde{LV}\}.$$

Here $A \subset V$ is a cycle and $\alpha \in \Gamma'$ makes it possible to index the copies of V contained in $\widetilde{LV}$. But the product on the Floer cohomology is obtained (by definition if we allow the possibility of setting $H'_i = 0$) by counting the triplets $(u(t)_\alpha, v(t)_\beta, w(t)_\gamma)$ belonging to $C_{A_\alpha} \times C_{B_\beta} \times C_{C_\gamma}$ for which there exists a pseudoholomorphic curve $\phi : \Sigma \to V$ such that

$$\phi_{|\partial D_1} = u, \quad \phi_{|\partial D_2} = v, \quad \phi_{|\partial D_3} = w, \quad \phi \# \tilde{u} \sqcup \tilde{v} = \tilde{w} \text{ in } \widetilde{LV}.$$

But such a curve, when completed by the disks $\tilde{u}, \tilde{v}, \tilde{w}$, provides precisely a pseudoholomorphic curve $\mathbb{P}^1 \to V$ of class a satisfying the condition

$$\omega(a) = \omega(\alpha) + \omega(\beta) - \omega(\gamma)$$

and intersecting the cycles A, B, C in fixed points of $\mathbb{P}^1$, so that the coefficient of $\exp\left(-\int_A \omega\right)$ in (5.39) is also equal to

$$\sum_{\bar{a}=\alpha+\beta-\gamma} \Phi_a\bigl([A], [B], [C]\bigr).$$

4. Equivariant cohomology

The equivariant cohomology of the complex projective space $\mathbb{P}^n$ endowed with the Hamiltonian action of S^1 is the principal tool used by Givental in [**91**]. Here we follow [**87**] and [**88**].

If G is a Lie group, we can construct a space BG that classifies principal G-bundles. The homotopy type of BG is characterized by the following property: *There exists a principal G-bundle $EG \to BG$ whose total space EG is contractible.*

We have the following result.

PROPOSITION 6.12. *If $E \to V$ is a principal G-bundle, there exists a continuous map $\phi : V \to BG$ defined up to homotopy such that $E \cong \phi^* EG$.*

Example. Let $G = S^1$ and

$$S^\infty = \varinjlim_n S^{2n+1}$$

where $S^{2n+1} \subset \mathbb{C}^{n+1}$ is the sphere having equation $\sum_i |z_i|^2 = 1$. It is easy to show that S^∞ is contractible. On the other hand, the compatible actions of $S^1 \subset \mathbb{C}^*$ on S^{2n+1} by coordinate multiplication provide a free action of S^1 on S^∞ whose quotient is

$$\mathbb{P}^\infty = \varinjlim_n \mathbb{P}^n.$$

Thus $\mathbb{P}^\infty$ is a model for BS^1. In particular, we have $H^*(BS^1, \mathbb{C}) = \mathbb{C}[h]$.

Let V be a variety endowed with a G-action. Then V is a retract of $V \times EG$ on which G acts diagonally, and this action is free. We set

$$V_G = (V \times EG)/G \text{ and } H^*_G(V) = H^*(V_G).$$

We have a natural map $\pi : V_G \to BG$ obtained by passing to the quotient of the projection $\mathrm{pr}_2 : V \times EG \to EG$, which endows the equivariant cohomology $H^*_G(V)$ with an $H^*(BG)$-module structure.

Example. If the action of G is trivial, we have $V_G = V \times BG$, whence $H^*_G(V) = H^*(V) \otimes H^*(BG)$. At the opposite extreme, if the action of G on V is free, V_G is a bundle having fiber EG over V/G, and hence $H^*_G(V) = H^*(V/G)$, the action of $H^*(BG)$ being trivial, that is, zero on $H^k(BG)$ for $k > 0$.

4.1. The equivariant de Rham cohomology.

Here we shall confine ourselves to describing the case where $G = S^1$, the general case being handled in [**87**]. Since BS^1 has an approximation by finite-dimensional manifolds, we can speak of differential forms on BS^1. The generator h of $H^*(BS^1)$ corresponding to the Euler class of the bundle $ES^1 \to BS^1$ can be represented by a differential form by choosing a connection θ on the principal bundle $ES^1 \to BS^1$ with group S^1 (which one can identify with a nonzero 1-form on ES^1 invariant under the action of S^1). The curvature form $d\theta$ is then a 2-form which is the pullback of a closed 2-form u on BS^1, which is a representative of the class h.

Now let V be a variety endowed with an action of S^1. Let X be the corresponding vector field on V, and let $\Omega_X^*(V)$ be the set of differential forms on V annihilated by the Lie derivative $\mathcal{L}_X$.

We consider $\Omega_X^*(V)[h]$ endowed with the grading

$$\deg_X \alpha = \deg \alpha, \quad \alpha \in \Omega_X^*(V), \quad \deg_X h = 2$$

and the differential d_X

(6.25) $$d_X(\alpha) = d\alpha + \mathrm{int}_X(\alpha) h, \quad d_X h = 0.$$

We then have (see [**87**]):

THEOREM 6.13. *There exists a natural isomorphism:*

(6.26) $$\mathrm{Ker}\left(d_X : \Omega_X^k(V)[h] \to \Omega_X^{k+1}(V)[h]\right) / \mathrm{Im}\left(d_X : \Omega_X^{k-1}(V)[h] \to \Omega_X^k(V)[h]\right)$$
$$\cong H_{S^1}^k(V, \mathbb{R}).$$

Specifically, let $\omega = \alpha + h\beta \in \Omega_X^*(V)[h]$. The condition $d_X(\omega) = 0$ is equivalent to

(6.27) $$d\alpha = 0, \quad \mathrm{int}_X(\beta) = 0, \quad d\beta = -\mathrm{int}_X(\alpha).$$

We then consider on $V \times ES^1$ the differential form

$$\widetilde{\omega} = \mathrm{pr}_1^* \alpha - d(\mathrm{pr}_2^* \theta \wedge \mathrm{pr}_1^* \beta).$$

This last is obviously closed and also satisfies

$$\mathrm{int}_{\widetilde{X}}(\widetilde{\omega}) = -\mathrm{pr}_1^* d\beta - \mathrm{int}_{\widetilde{X}}(\mathrm{pr}_2^*(\widetilde{u}) \wedge \mathrm{pr}_1^*(\beta) - \mathrm{pr}_2^*\theta \wedge \mathrm{pr}_1^* d\beta),$$

where $\widetilde{u} = d\theta$ is the *pull-back* on ES^1 of the form u on BS^1 and $\widetilde{X}$ is the field associated with the diagonal action of S^1 on $V \times ES^1$. But the right-hand side is zero since $\mathcal{L}_X \beta = 0$, $\mathrm{int}_X \beta = 0$ and $\mathrm{int}_{\widetilde{X}}(\widetilde{u}) = 0$, which implies

$$\mathrm{int}_{\widetilde{X}}(\mathrm{pr}_2^*(\widetilde{u}) \wedge \mathrm{pr}_1^*(\beta) - \mathrm{pr}_2^*\theta \wedge \mathrm{pr}_1^* d\beta) = -\mathrm{pr}_1^* d\beta.$$

This form thus projects down as a closed form on V_G. One can carry out a similar computation for the polynomials of any degree in h, which shows how to construct the isomorphism (6.26).

In the rest of this section we discuss some facts relative to the S^1-equivariant cohomology of complex projective space, where S^1 acts linearly. These facts are applied in the following sections to the space of Laurent polynomials $P(z)$ with coefficients in $\mathbb{C}^{n+1}$ endowed with the action $\lambda \cdot P(z) = P(\lambda z)$ for $\lambda \in S^1$.

4.2. The case of a Hamiltonian action.
Let (V, ω) be a symplectic manifold endowed with a Hamiltonian action of S^1: $\text{int}_X(\omega) = dH$. We then have

LEMMA 6.14. *There exists a class $p \in H^2_{S^1}(V)$ whose image in $H^2(V)$ under the natural restriction is equal to the cohomology class of ω.*

PROOF. We use the description given in 4.1 for $H^2_{S^1}(V)$. We have
$$d\omega = 0, \quad \mathcal{L}_X \omega = 0, \quad \mathcal{L}_X H = 0, \quad dH = \text{int}_X(\omega),$$
so that $(\omega - Hh)$ is in $\Omega^2_X(V)[h]$ and is d_X-closed. It thus provides an element p of $H^2_{S^1}(V)$ represented by the form $\text{pr}_1^*(\omega) + d(\text{pr}_1^* H \text{pr}_2^* \theta)$, which can be regarded as a form on V_G. This form clearly has ω as its restriction to each fiber of the map $\pi : V_{S^1} \to B_{S^1}$. □

4.3. The case of projective space.
For $i_0, \ldots, i_N \in \mathbb{Z}$ we consider the action of S^1 on $\mathbb{P}^N$ defined by
$$(6.28) \qquad \lambda \cdot (x_0, \ldots, x_N) = (\lambda^{i_0} x_0, \ldots, \lambda^{i_N} x_N).$$

This action preserves the symplectic form
$$\omega = \frac{i}{2} \sum_j dx_j \wedge d\bar{x}_j, \quad \sum_j |x_j|^2 = 1,$$

and the corresponding Hamiltonian function is
$$H = \sum_k i_k |x_k|^2, \quad \sum_j |x_j|^2 = 1.$$

By the proof of Lemma 6.14, we thus have on $\mathbb{P}^N_{S^1}$ a form
$$\tilde{\omega} = \text{pr}_1^* \omega + d(\text{pr}_1^* H \text{pr}_2^* \theta)$$
whose restriction to each fiber (isomorphic to $\mathbb{P}^N$) of $\pi : \mathbb{P}^N_{S^1} \to B_{S^1}$ generates the cohomology of $\mathbb{P}^N$. As above, we use p to denote its class in $H^*_{S^1}(\mathbb{P}^N, \mathbb{C})$.

LEMMA 6.15. *Let $H^*_{S^1}(\mathbb{P}^N, \mathbb{C})_0$ be the localization of the $\mathbb{C}[h]$-module $H^*_{S^1}(\mathbb{P}^n, \mathbb{C})$ obtained by inverting h. Then there exists a natural isomorphism*
$$(6.29) \qquad H^*_{S^1}(\mathbb{P}^N, \mathbb{C})_0 \cong \mathbb{C}[p, h, h^{-1}]/(p - i_0 h) \cdots (p - i_N h).$$

PROOF. It follows from the Leray–Hirsch theorem and the fact that the restriction of p generates the cohomology of the fibers of the map π that we have a surjection $f : \mathbb{C}[p, h] \to H^*_{S^1}(\mathbb{P}^N, \mathbb{C})$. It thus suffices to determine the kernel of the localization of f.

We now note that if $\mathbb{P}_k$ is the set of fixed points of the action (6.28) made up of the points $(x_0, \ldots, x_N)$ such that $x_j = 0$ for $i_j \neq k$ and if n_k is the dimension of $\mathbb{P}_k$, then H equals k on $\mathbb{P}_k$, so that we have the relation $\tilde{\omega} - kh_{|\mathbb{P}_k} = \omega_{|\mathbb{P}_k}$ in $H^*_{S^1}(\mathbb{P}_k) \cong H^*(\mathbb{P}_k)[h]$, since $d\theta$ is the pullback of a form of class h in $H^*(BS^1)$. From that we deduce that
$$\prod_{\{k | \mathbb{P}_k \neq \emptyset\}} (p - kh)^{n_k + 1}$$

has a zero restriction in the cohomology of the locus of the fixed points of the action (6.28). According to the localization theorem [88], we know that the restriction map

$$j^* : H^*_{S^1}(\mathbb{P}^N, \mathbb{C})_0 \to \bigoplus H^*_{S^1}(\mathbb{P}_k, \mathbb{C})_0$$

is injective (indeed, an isomorphism). From it we deduce that the kernel of the map

$$f : \mathbb{C}[p, h, h^{-1}] \to H^*_{S^1}(\mathbb{P}^N, \mathbb{C})_0$$

is generated by

$$\prod_{\{k | \mathbb{P}_k \neq 0\}} (p - kh)^{n_k+1},$$

which proves that

$$H^*_{S^1}(\mathbb{P}^n, \mathbb{C})_0 \cong \mathbb{C}[p, h, h^{-1}]/(p - i h_0) \cdots (p - i_N h).$$

$\square$

4.4. Integration on the fiber. Let V be a compact variety endowed with an action of S^1. The map $\pi : V_{S^1} \to BS^1$ has V as fiber, and so we have an integration map on the fiber

$$\pi_* : H^*_{S^1}(V) \longrightarrow H^{*-\dim V}(BS^1)$$

which is a morphism of $H^*(BS^1)$-modules and can thus be localized. Let V_f be the locus of the fixed points of the action of S^1 and $j : V_f \hookrightarrow V$ the inclusion. The localization isomorphism

(6.30) $$H^*_{S^1}(V, \mathbb{C})_0 \overset{j^*}{\cong} H^*_{S^1}(V_f, \mathbb{C}) \cong H^*(V_f) \otimes \mathbb{C}[h, h^{-1}]$$

makes it possible to express the map π_* conveniently.

For each component V_f^α of V_f of codimension m_α, let $j_\alpha = j_{|V_f^\alpha}$. We then have the Gysin morphism

$$j_{\alpha*} : H^*_{S^1}(V_f^\alpha) \to H^{*+m_\alpha}_{S^1}(V)$$

obtained as the composition

(6.31) $$H^*(V_f^\alpha \times BS^1) \overset{\text{Thom}}{\underset{\sim}{=\!=\!=}} H^{*+m_\alpha}(V_{S^1}, V_{S^1} - V_f^\alpha \times BS^1) \to H^{*+m_\alpha}(V_{S^1}).$$

It satisfies the following property. Let

$$\pi_\alpha = \pi_{|V_F^\alpha \times BS^1}$$

(the second projection). Then

$$\pi_{\alpha*} = \pi_* \circ j_{\alpha*}$$

and in addition

$$j_\alpha^* \circ j_{\alpha*} = \begin{cases} \text{cup-product with the equivariant Euler class } e_\alpha \in H^*(V_f^\alpha \times BS^1) \\ \text{of the normal bundle } N_\alpha \text{ of } V_f^\alpha \text{ in } V, \text{ endowed with the action of } S^1 \\ \text{given by the differential } \lambda_* \text{ of } \lambda \in S^1 \subset \text{Diff}(V). \end{cases}$$

4. EQUIVARIANT COHOMOLOGY

This equivariant Euler class is simply the usual Euler class of the bundle $\widetilde{N}_\alpha$ on $V_f^\alpha \times BS^1$ obtained as the quotient of the bundle $\mathrm{pr}_1^*(N_\alpha)$ over $V_f^\alpha \times ES^1$ modulo the action of S^1

$$\lambda \cdot (n, v, x) = \bigl(\lambda_*(v)(n), v, \lambda \cdot x\bigr).$$

But the localization $e_\alpha \in H^*_{S^1}(V_f^\alpha)_0 = H^*(V_f^\alpha) \otimes \mathbb{C}[h, h^{-1}]$ is invertible, by the fact that its term of degree 0 in $H^*(V_f^\alpha)$ is the Euler class $e \in H^*(BS^1)$ of the bundle $\widetilde{N}_{\alpha|v \times BS^1}$ for all $v \in V_f^\alpha$. Since the action of S^1 on $N_{\alpha,v}$ has no fixed points except 0, $N_{\alpha,v}$ splits (for example, by using an S^1-invariant metric on V) into a direct sum of spaces L^i_α of dimension 2 endowed with a complex structure, on which $\lambda_*(v)$ can be identified with the multiplication by λ^{k_i} with $k_i \neq 0$. The Euler class of the corresponding bundle $\widetilde{L}^i_\alpha$ on BS^1 is then equal to $k_i h$, and we thus have

$$e = \prod_i k_i h^{m_\alpha/2},$$

which is invertible in $\mathbb{C}[h, h^{-1}]$. This implies the invertibility of e_α immediately.

REMARK 6.16. This shows that the localized map j^* has a right inverse $\bigoplus_\alpha j_{\alpha*} \circ (1/e_\alpha \wedge)$, and hence that the localization arrow (6.30) is surjective.

The preceding shows Bott's formula:

$$(6.32) \qquad \pi_* \eta = \sum_\alpha \int_{V_f^\alpha} j_\alpha^* \eta / e_\alpha \in \mathbb{C}[h, h^{-1}] \text{ for } \eta \in H^*_{S^1}(V),$$

where we have identified $\pi_{\alpha*} : H^*(V_f^\alpha) \otimes \mathbb{C}[h, h^{-1}] \to \mathbb{C}[h, h^{-1}]$ with the usual integration on V_f^α.

In particular, in the case considered in § 4.3, we have the following description of π_*.

PROPOSITION 6.17. *With the notation of § 4.3, the integration arrow on the fiber composed with the isomorphism (6.29) is given by*

$$(6.33) \qquad \pi_*\bigl(f(p,h)\bigr) = \text{sum of the residues of } f(p,h) \Big/ \prod_l (p - i_l h)\, dp.$$

PROOF. In fact this results from Bott's formula, the components V_f^α in this case being the spaces $\mathbb{P}_k$ with real codimension $2(N - n_k)$, and the normal bundle of $\mathbb{P}_k$ in $\mathbb{P}^N$ being isomorphic to $\mathcal{O}_{\mathbb{P}_k}(1) \otimes T^{N-n_k}$, where the action of S^1 on the trivial bundle T^{N-n_k} with basis $\partial/\partial x_j$ for $i_j \neq k$ is given by

$$(6.34) \qquad \lambda \cdot \Bigl(\frac{\partial}{\partial x_j}\Bigr)_{i_j \neq k} = \Bigl(\lambda^{k-i_j} \frac{\partial}{\partial x_j}\Bigr)_{i_j \neq k}.$$

The bundle $\widetilde{N}_k$ on $\mathbb{P}^k \times BS^1$ thus admits as Euler class

$$\prod_{\{j \mid i_j \neq k\}} (\omega_{|\mathbb{P}_k} + (k - i_j)h)$$

and we thus have, for $\eta = f(p, h) \in H^*_{S^1}(\mathbb{P}^N)_0$

$$(6.35) \qquad \pi_*(\eta) = \sum_{\{k \mid \mathbb{P}_k \neq \emptyset\}} \int_{\mathbb{P}_k} j_k^*\bigl(f(p,h)\bigr) \Big/ \prod_{\{j \mid i_j \neq k\}} (\omega_{|\mathbb{P}_k} + (k - i_j)h),$$

where $j_k^*(p - kh) = \omega_{|\mathbb{P}_k}$. But it can be verified immediately that
$$\int_{\mathbb{P}_k} \frac{j_k^*(f(p,h))}{\prod_{\{j|i_j \neq k\}} (\omega_{|\mathbb{P}_k} + (k - i_j)h)}$$
is equal to the residue at $p = kh$ of the differential form
$$\frac{f(p,h)}{\prod_l (p - i_l h)} dp.$$

Indeed, these two quantities are equal to the coefficient of ω^{n_k} in the expansion of $f(p,h)/\prod_l (p - i_l h)$ in powers of $\omega = p - kh$. Thus $\pi_*(\eta)$ is the sum of the residues of $f(p,h)/\prod_l (p - i_l h) dp$. $\square$

5. The Givental construction

5.1. The $\mathcal{D}$-module structure on the equivariant cohomology. The construction we are about to describe here is an essential ingredient of [**91**]. It explains the sense in which an equivariant homology class can be a solution of a differential equation.

Let (V, ω) be a symplectic manifold endowed with a locally Hamiltonian S^1-action. Assume that the closed 1-form $\text{int}_X(\omega)$ is of primitive integer class. Then S^1 acts on the the covering $\widetilde{V} \to V$ with group $\mathbb{Z}$ defined by the kernel of the map $\pi_1(V) \to \mathbb{Z}$, $l \mapsto \int_l \text{int}_X(\omega)$, since $\text{int}_X(\omega)$ has zero integral over the orbits of S^1, and the induced action of S^1 on $\widetilde{V}$ is Hamiltonian. The function H on $\widetilde{V}$ such that $dH = \text{int}_X(\omega)$ satisfies $q^*H = H - 1$, where q is the action of the generator l of the group of the covering such that $\int_l \text{int}_X(\omega) = -1$. (We also use ω to denote the symplectic form induced on $\widetilde{V}$.)

On $\widetilde{V} \times ES^1$, let
$$\widetilde{\omega} = \text{pr}_1^* \omega + d(\text{pr}_1^* H \text{pr}_2^* \theta)$$
be the associated form, descending toto V_{S^1} and providing an equivariant cohomology class $p \in H_{S^1}^*(V)$ (see 4.2). We have $q^*(\widetilde{\omega}) = \widetilde{\omega} - d(\text{pr}_2^*\theta)$ on $\widetilde{V} \times ES^1$, whence the identity

(6.36) $$q^*(p) = p - h \in H_{S^1}^*(\widetilde{V}),$$

since $d\theta$ has π^*h, which we also denote ∂yh, as its class in $H_{S^1}^*(\widetilde{V})$. On the other hand, we evidently have $q^*h = h$. If we regard q^* and p as operators acting on $H_{S^1}^*(\widetilde{V})$ (the second by cup-product with the class p), we thus have:

(6.37) $$\forall \alpha \in H_{S^1}^*(\widetilde{V}), \quad (p \circ q^* - q^* \circ p)(\alpha) = hq^*\alpha,$$

that is, the commutation relation
$$p \circ q^* - q^* \circ p = hq^*$$
in the ring of endomorphisms of $H_{S^1}^*(\widetilde{V})$. We note that this relation is exactly the relation between the operators $p' = h\partial/\partial t$ and $q' = e^t \times$ acting on functions of t. (Here h is a scalar or a formal variable.) Let $\mathcal{D}$ be the ring of differential operators on the circle generated by p' and q'. We have constructed a $\mathcal{D}$-module structure on $H_{S^1}^*(\widetilde{V})$.

5. THE GIVENTAL CONSTRUCTION

5.2. Application to the loop space. Let (V, ω) be a symplectic variety and LV the space of contractible loops in X. To simplify assume that ω is of integer class and that the map $\omega : \pi_2(V) \to \mathbb{Z}$ is onto. We then have a covering $\widetilde{LV} \to LV$ with group $\mathbb{Z}$ defined as in § 1, and we can define the following free action on $\widetilde{LV}$ (see (6.2))):

$$\mathcal{A}(\tilde{\phi}) = \int_{D^2} \tilde{\phi}^* \omega. \tag{6.38}$$

The space LV (resp. $\widetilde{LV}$) is endowed with a symplectic structure

$$\omega_{LV}\big(\xi(t), \chi(t)\big) = \int_{S^1} \omega\big(\xi(t), \chi(t)\big)\, dt$$

for $\phi \in LV$, and tangent vectors $\xi(t), \chi(t) \in \mathcal{C}^\infty(\phi^* T_V)$ to LV at ϕ. The circle S^1 operates on LV (resp. $\widetilde{LV}$) by rotation of the loops: $\lambda \cdot \phi(z) = \phi(\lambda z)$. We have

$$d\mathcal{A}(\xi) = -\int_{S^1} \omega\big(\phi'(t), \xi(t)\big)\, dt,$$

so that $\mathcal{A}$ is up to sign the Hamiltonian function corresponding to the action of S^1 on $\widetilde{LV}$. Let

$$q(\tilde{\phi}) = \alpha \# \tilde{\phi}$$

be the action of the generator α of the covering group $\pi_2/\mathrm{Ker}\,(\omega) \cong \mathbb{Z}$ such that $\int_\alpha \omega = 1$. We have

$$q^* \mathcal{A} = \mathcal{A} + 1,$$

so that we have the data described in 5.1.

Givental wishes to apply the construction of a $\mathcal{D}$-module structure (see 5.1) not to the S^1-equivariant cohomology of $\widetilde{LV}$ but to the "Floer S^1-equivariant cohomology" of $\widetilde{LV}$, which is unfortunately not defined. Combining the isomorphism (6.16) and the localization isomorphism (6.30), he sets

$$FH^*_{S^1}(\widetilde{LV}, \mathbb{C}) = H^*(V) \otimes \mathbb{C}[h, h^{-1}] \otimes \Lambda_q, \tag{6.39}$$

where Λ_q is the ring of formal Laurent series $\sum_{k \geq N} a_k q^k$, $a_k \in \mathbb{C}$, $N \in \mathbb{Z}$. (The fixed points of the action of S^1 on $\widetilde{LV}$ are different copies of V contained in $\widetilde{LV}$.) Givental assumes that $FH^*_{S^1}(\widetilde{LV}, \mathbb{C})$ can be calculated using the cycles C_{A_α} considered in Remark 6.8. He uses this only as a heuristic argument to help interpret the calculations he is performing.

5.3. The appproximation of $\widetilde{LP^n}$. We now study the case when V is the projective space $\mathbb{P}^n$ endowed with its form ω (see 4.3). For each $k \in \mathbb{N}$ we consider the projective space of Laurent polynomials

$$M_k = \left\{ \sum_{-k \leq l \leq k} \phi_l z^l;\ \phi_l \in \mathbb{C}^{n+1},\ \phi_l \text{ not all zero} \right\} \Big/ \mathbb{C}^*.$$

Let

$$M = \varinjlim_n M_k.$$

Then M admits an action of $\mathbb{Z}$ without fixed points:
$$q \cdot \left(\sum_{-k \leq l \leq k} \phi_l z^l \right) = \sum_{-k \leq l \leq k} \phi_l z^{l+1}.$$

The quotient of M by $\mathbb{Z}$ parametrizes the algebraic loops in $\mathbb{P}^n$, which are dense in $L\mathbb{P}^n$, so that M is an approximation of $\widetilde{L\mathbb{P}^n}$. On the other hand, M is endowed with the action of S^1 given by rotation of loops as in § 5.2.

Givental then works essentially with the equivariant cohomology of M_k. The action of S^1 on M_k is described in the coordinates $x_i^l = \phi_{l,i}$ for $0 \leq i \leq n$ and $-k \leq l \leq k$, where the x_i are coordinates for $\mathbb{P}^n$, by
$$\lambda \cdot (x_i^l) = (\lambda^l x_i^l).$$

By 4.3 we thus have
$$(6.40) \qquad H^*_{S^1}(M_k, \mathbb{C})_0 = \mathbb{C}[p, h, h^{-1}] \bigg/ \prod_{-k \leq l \leq k} (p + lh)^{n+1}.$$

We note that the class p on M_k restricts to the class p on M_{k-1} and is the equivariant Euler class of the equivariant bundle $\mathcal{L}$ of rank 1 dual to the tautologic bundle whose fiber at the point (x_i^l) is the line generated by (x_i^l) endowed with the action of S^1 given by
$$\lambda \cdot \left((x_i^l), (\alpha x_i^l) \right) = \left((\lambda^l x_i^l), (\alpha \lambda^l x_i^l) \right).$$

The map $q : M_k \to M_{k+1}$ satisfies
$$q^* \mathcal{L} \cong \mathcal{L} \otimes \mathcal{T}_1,$$
where $\mathcal{T}_i$ is the trivial bundle of rank 1 on M_k endowed with the following action of S^1:
$$\lambda \cdot (x, \alpha) = (\lambda \cdot x, \lambda^{-i} \alpha).$$

The equivariant Euler class of $\mathcal{T}_1$ is equal to $-h$. We thus have
$$q^* p = p - h.$$

Let $M_k^+ \subset M_k$ be the projective space of *polynomials*
$$M_k^+ = \left\{ \sum_{0 \leq l \leq k} \phi_l z^l \right\} \bigg/ \mathbb{C}^*.$$

This space is invariant under the action of S^1 and admits an equivariant cohomology class $\Delta_k \in H^*_{S^1}(M_k)$, which is simply defined as the cohomology class of $(M_k^+)_{S^1} \subset (M_k)_{S^1}$.

LEMMA 6.18. *The image of the class Δ_k under the isomorphism (6.40) is equal to $\prod_{0 < l \leq k} (p + lh)^{n+1}$ (modulo the relation $\prod_{-k \leq l \leq k} (p + lh)^{n+1}$).*

Indeed, M_k^+ is the complete intersection of the hyperplanes having equation
$$x_i^l = 0 \quad \text{for } 0 \leq i \leq n \text{ and } l < 0.$$
But x_i^l is an invariant section under S^1 of the bundle $\mathcal{L} \otimes \mathcal{T}_l$. Since the bundle $\mathcal{T}_l$ has Euler class $-lh$, we deduce from this, as stated, that
$$(6.41) \qquad \Delta_k = \prod_{-k \leq l < 0} (p - lh)^{n+1} = \prod_{0 < l \leq k} (p + lh)^{n+1}. \qquad \square$$

We note that
$$\varinjlim_k M_k^+ =: M^+ \subset M$$
is essentially the "fundamental Floer homology cycle" $C_{\mathbb{P}^n}$ consisting of the polynomial maps $\phi : D^2 \to \mathbb{P}^n$. Givental thus sets

(6.42) $$\Delta = \lim_{k \to \infty} \Delta_k = \prod_{l>0}(p + lh)^{n+1},$$

which unfortunately has no meaning even in $H_{S^1}^*(\mathbb{P}^n) \otimes \Lambda_q$, which is a reasonable completion of $\varinjlim_k H_{S^1}^*(M_k)$. (Here $\mathbb{P}^n$ is endowed with the trivial action of S^1 and is regarded as the locus of the fixed points of the action of S^1 on $L\mathbb{P}^n$.) Since we have $q^*p = p - h$, this infinite product satisfies formally the equation

(6.43) $$q^*\Delta = p^{n+1}\Delta.$$

5.4. The smooth hypersurfaces of P^n. We now conclude the calculation of [91]. We shall consider the S^1-equivariant cohomology class in M_k, $k \to \infty$, of the set of polynomials $P(z)$ of degree k with values in a hypersurface $X \subset \mathbb{P}^n$ and calculate the differential equation that it satisfies relative to the $\mathcal{D}$-module structure defined in 5.1 and 5.2.

Let $X \subset \mathbb{P}^n$ be a smooth hypersurface defined by a homogeneous polynomial F of degree d. Consider the subset

$$M_k^+(X) = \{\Phi \in M_k^+;\ F(\Phi(x)) = 0\} \subset M_k^+.$$

This is the locus of the zeros of the S^1-invariant section $\Phi \mapsto F(\Phi)$ of the equivariant bundle

$$\mathcal{L}^d \otimes E \longrightarrow M_k^+,$$

where E is the trivial bundle with fiber equal to the set of polynomials in z of degree at most kd endowed with the action of S^1 defined by $\lambda \cdot z^l = \lambda^{-l}z^l$. Even though $M_k^+(X)$ is practically never of the correct codimension

$$kd + 1 = \operatorname{rank} E,$$

it is natural to estimate that the interesting class is the equivariant Euler class Δ_k^d of $\mathcal{L}^d \otimes E$ rather than that of $M_k^+ X)$. The equivariant bundle E on M_k being the direct sum of the bundles $\mathfrak{T}_l$ for $0 \leq l \leq kd$, we thus obtain:

(6.44) $$\Delta_k^d = \prod_{0 \leq l \leq kd}(dp - lh) \in H_{S^1}^*(M_k^+).$$

Finally, since $j_k : M_k^+ \hookrightarrow M_k$ has equivariant cohomology class

$$\Delta_k = \prod_{0 < l \leq k}(p + lh)^{n+1},$$

we find

(6.45) $$j_{k*}(\Delta_k^d) = \prod_{0 < l \leq k}(p + lh)^{n+1} \prod_{0 \leq l \leq kd}(dp - lh) \in H_{S^1}^*(M_k),$$

an expression we shall denote by $\Delta_k \cdot \Delta_k^d$.

Consider the infinite product
$$\Delta \cdot \Delta^d = \prod_{l>0}(p+lh)^{n+1} \prod_{l\geq 0}(dp-lh)$$
(which should represent the "fundamental Floer cycle of X in M," but actually has no meaning). Since $q^*(p) = p - h$, we find that $\Delta \cdot \Delta^d$ satisfies formally the equation

(6.46) $\quad q^*\big(d(dp+h)(dp+2h)\cdots(dp+(d-1)h)\Delta \cdot \Delta^d\big)$
$$= d\big(dp + (1-d)h\big)\big(dp + (2-d)h\big)\cdots(dp - h)$$
$$\times \prod_{l\geq 0}(p+lh)^{n+1} \prod_{l\geq 0}\big(dp - (d+l)h\big)$$
$$= \frac{dp^{n+1}}{dp}\Delta \cdot \Delta^d = p^n \Delta \cdot \Delta^d.$$

Replacing p by $h\partial/\partial t$, then q^* by e^t and $\Delta \cdot \Delta^d$ by $f(t)$, we convert this equation into

(6.47) $\quad h^{n-d+1}\left(\dfrac{\partial}{\partial t}\right)^n f = e^t d\left(d\dfrac{\partial}{\partial t}+1\right)\left(d\dfrac{\partial}{\partial t}+2\right)\cdots\left(d\dfrac{\partial}{\partial t}+d-1\right)f.$

The remarkable point in this construction is the fact that for $n = 4$ and $d = 5$, this equation is exactly the Picard–Fuchs equation of the mirror family of the family of quintics (see [**43**] and § 4), calculated using the form ω'_ψ defined in Remark 3.15, for the logarithmic vector field $\psi'\partial/\partial\psi'$, $\psi' = 1/\lambda^5 = e^t$ (so that $\psi'\partial/\partial\psi' = \partial/\partial t$).

Since the class $\Delta \cdot \Delta_d$ is the virtual class of $M_k^+(X)$ and is thus essentially defined using the rational curves on a hypersurface X of degree d in $\mathbb{P}^n$, we can consider this fact as a justification of the identification of the two series of §3.

REMARK 6.19. Another interesting point is the disappearance of the variable h in the equation (6.47) precisely when $d = n+1$, that is, when $X \subset \mathbb{P}^n$ has trivial canonical bundle.

Finally, Givental constructs formal solutions of this equation as follows. Let $q_* = (q^{-1})^*$ be the adjoint operator of q. For a cohomology class C in $H^*_{S^1}(M)$, we have (formally):

(6.48) $\quad\quad\quad\quad\quad\quad \pi_*(q_*C \cdot \Delta \cdot \Delta^d) = \pi_*(C \cdot q^*\Delta \cdot \Delta^d).$

Assume that C satisfies the condition
$$q_*C = C.$$
Then we have, for $\Gamma_C = e^{pt/h}C$,
$$q_*(\Gamma_C) = e^t \Gamma_C$$
(since $q_*p = p + h$) and
$$p \cdot \Gamma_C = h\dfrac{\partial}{\partial t}\Gamma_C.$$
From this we deduce that $f_h(t) = \pi_*(\Gamma_C \cdot \Delta \cdot \Delta^d)$ satisfies

(6.49) $\quad \begin{cases} h\dfrac{\partial f_h}{\partial t} = \pi_*(\Gamma_C \cdot p \cdot \Delta \cdot \Delta^d), \\ e^t f_h = \pi_*(e^t \Gamma_C \cdot \Delta \cdot \Delta^d) = \pi_*(q_*\Gamma_C \cdot \Delta \cdot \Delta^d) = \pi_*(\Gamma_C \cdot q^*\Delta \cdot \Delta^d). \end{cases}$

It follows immediately that $f_h(t)$ is a solution of the differential equation (6.47).

Here π_* is calculated by passing to the limit as $k \to \infty$ in formula (6.33) which gives $\pi_{k*}f(p,h)$ as a sum of residues of

$$\frac{f(p,h)}{\prod_{-k \leq l \leq k}(p+lh)}\,dp.$$

Bibliography

[1] R. Bryant, P. Griffiths, *Some observations on the infinitesimal period relations for regular threefolds with trivial canonical bundle*, in: *Arithmetic and Geometry* (papers dedicated to Shafarevich), Vol. 2, Birkhäuser, 1983.

[2] A. Beauville, *Variétés kählériennes dont la première class de Chern est nulle*, J. Diff. Geom., **18** (1983), 755–782.

[3] F. A. Bogomolov, *On the decomposition of Kähler manifolds with trivial canonical class*, Math. USSR, Sbornik, **22** (1974), 580–583.

[4] F. A. Bogomolov, *Hamiltonian Kähler manifolds*, Soviet Math. Doklady, **19** (1978), 1462–1465.

[5] C. Borcea, *K3-surfaces with involution and mirror pairs of Calabi–Yau manifolds*, preprint (1992).

[6] J. P. Demailly, *Théorie de Hodge L^2 et théorèmes d'annulation*, in [**25**].

[7] R. Donagi, E. Markman, *Cubics, integrable systems, and Calabi–Yau threefolds*, in: *Proceedings of the Hirzebruch 65 Conf. on Algebraic Geometry*, Israel Math. Conf. Proc., **9** (1996).

[8] R. Friedman, *On threefolds with trivial canonical bundle*, Complex Geometry and Lie Theory (Sundance, UT, 1989), Proc. Symp. Pure Math. **53** (1991), 103–134.

[9] R. Friedman, *Global smoothing of varieties with normal crossings*, Annals of Math. **118** (1983), 85–114.

[10] P. Griffiths, *Periods of integrals on algebraic manifolds I and II*, Amer. J. Math. **90** (1968), 568–626 and 805–865.

[11] Y. Kawamata and Y. Namikawa, *Logarithmic deformations of normal crossing varieties and smoothing of degenerate Calabi–Yau varieties*, Invent. Math. **118** (1994), 395–409.

[12] M. Kontsevich, *Homological algebra of Mirror Symmetry*, Proceedings du Congrès International de Zürich, Vol. 1, 101–139, Birkhäuser, 1995.

[13] D. Markuschevich, *Resolution of singularities* (appendix), Commun. Math. Phys., **111** (1987), 247–274.

[14] V. V. Nikulin, *Discrete reflection groups in Lobachevski spaces and algebraic surfaces*, Proceedings of the International Congress of Mathematicians, Berkeley, 1986, 654–671.

[15] Z. Ran, *Deformation of manifolds with torsion or negative canonical bundle*, J. Alg. Geom., **1** (1992), No. 2, 279–291.

[16] S. S. Roan, *On Calabi–Yau orbifolds in weighted projective space*, Int. J. Math., **1** (1990), 211–232.

[17] S. S. Roan, *Mirror symmetry and Arnold's duality*, preprint MPI/92–86, Max Planck Institut (1992).

[18] G. Tian, *Smoothness of the universal deformation space of compact Calabi–Yau manifolds and its Peterson–Weil metric*, in: *Mathematical Aspects of String Theory*, S. T. Yau, ed., World Scientific Press, Singapore (1987), 629–646.

[19] M. Verbitsky, *Mirror symmetry for hyperkähler manifolds*, preprint (1995).

[20] C. Voisin, *Miroirs et involutions sur les surfaces K3*, in: *Journées de géométrie algébrique d'Orsay, juillet 1992*, A. Beauville, O. Debarre, and Y. Laszlo, eds., Astérisque, **218** (1993).

[21] A. Weil, *Variétés kählériennes*, Hermann, Paris, 1957.

[22] P. H. M. Wilson, *The Kähler cone on Calabi–Yau threefolds*, Invent. Math., **107** (1992), 561–583.

[23] *Première classe de Chern et courbure de Ricci: preuve de la conjecture de Calabi*, Séminaire Palaiseau, Astérisque, **58** (1978).

[24] *Géométrie des surfaces K3: modules et périodes*, Séminaire Palaiseau, Astérisque, **126** (1985).

[25] J. Bertin, J. P. Demailly, L. Illusie, and C. Peters, *Introduction à la théorie de Hodge*, États de la recherche, Grenoble 1992, Panoramas et synthèses, **3** (1996).

Chapter 2

[26] L. Alvarez-Gaumé and D. Freedman, *Geometrical structure and ultraviolet finiteness in the supersymmetric σ-model*, Commun. Math. Phys., **80** (1981), 443–451.

[27] P. Aspinwall, C. Lütken, and G. Ross, *Construction and couplings of mirror manifolds*, Phys. Lett. B, **244** (1990), No. 3.

[28] J. B. Bost, *Fibrés déterminants, déterminants régularisés et mesure sur les espaces de modules de courbes complexes*, Séminaire Bourbaki, exp. 676 (1987), Astérisque, 152–153.

[29] K. Gawedzki, *Conformal field theory*, Séminaire Bourbaki, exp. 704, 41st year, 1988–1989.

[30] D. Gepner, *Exactly solvable string compactifications on manifolds of $SU(n)$ holonomy*, Phys. Lett. B, **199** (1987), No. 3, 380–388.

[31] B. R. Greene and M. R. Plesser, *Duality in Calabi–Yau moduli spaces*, Nucl. Phys. B, **338** (1988), 15–37.

[32] B. R. Greene, D. R. Morrison, and M. R. Plesser, *Mirror manifolds in higher dimension*, Commun. Math. Phys., **173** (1995), 559–598.

[33] N. Hitchin, A. Karlhede, U. Lindström, and M. Roček, *Hyperkähler metrics and supersymmetry*, Commun. Math. Phys., **108** (1987).

[34] H. B. Lawson, and M. L. Michelsohn, *Spin Geometry*, Princeton Mathematical Series, Princeton University Press, New Jersey, 1989.

[35] W. Lerche, C. Vafa, and N. P. Warner, *Chiral rings in $N=2$-superconformal field theories*, Nucl. Phys. B, **324** (1989), 427–474.

[36] G. Segal, *The definition of conformal field theory*, course notes.

[37] G. Segal, *The definitions of conformal field theory*, in: Links between Geometry and Mathematical Physics, 13–17.

[38] E. Witten, *Mirror manifolds and topological field theory*, in [**39**], 120–158.

[39] *Essays on Mirror Manifolds*, S. T. Yau, ed., International Series in Mathematical Physics, International Press, 1992.

[40] B. Dewitt, *Supermanifolds*, Cambridge Monographs on Mathematical Physics, 1984.

Chapter 3

[41] V. Batyrev and D. van Straten, *Generalized hypergeometric functions and rational curves on Calabi–Yau complete intersections in toric varieties*, preprint.

[42] J. Bertin and C. Peters, *Variations de structure de Hodge, variétés de Calabi–Yau et symétrie miroir*, in [**25**].

[43] P. Candelas, X. C. de la Ossa, P. S. Green, and L. Parkes, *A pair of Calabi–Yau manifolds as an exactly soluble superconformal field theory*, Nucl. Phys. B, **359** (1991), 21–74.

[44] J. Carlson and P. Griffiths, *Infinitesimal variations of Hodge structures and the global Torelli problem*, in: Géométrie algébrique, Angers, A. Beauville, ed., Sijthoff–Noordhoff, 1980, 51–76.

[45] G. Ellingsrud and S. A. Stromme, *The number of twisted cubics on the general quintic threefold*, in [**39**].

[46] G. Ellingsrud and S. A. Stromme, *Bott's formula and enumerative geometry*, J. Amer. Math. Soc., **9** (1996), 175–193.

[47] P. Griffiths, *On the periods of certain rational integrals, I, II*, Ann. Math., **90** (1969), 460–541.

[48] P. Griffiths and L. Tu, *Curvature properties of Hodge bundles*, in [**59**], 29–49.

[49] A. Grothendieck, *On the de Rham cohomology of algebraic varieties*, Publ. Math. I.H.E.S., **29** (1966), 96–103.

[50] N. Katz, *The regularity theorem in algebraic geometry*, Actes du congrès international des mathématiciens, Nice 1970, Vol. 1, 437–443.

[51] M. Kontsevich, *Enumeration of rational curves via torus action*, in: The moduli space of curves, R. Dijkgraaf, C. Faber, G. van der Geer, eds., Progress in Math., **129**, Birkhäuser, 1995, 335–368.

[52] A. Libgober and J. Teitelbaum, *Lines on Calabi–Yau complete intersections, mirror symmetry, and Picard–Fuchs equations*, International Research Notices, No. 1, 1993.

[53] D. Morrison, *Mirror symmetry and rational curves on quintic threefolds*, J. Amer. Math. Soc., **6**, No. 1, 223–241.

[54] D. Morrison, *Picard–Fuchs equations and mirror maps for hypersurfaces*, in [**39**], 241–264.

[55] D. Morrison, *Compactifcations of moduli spaces inspired by mirror symmetry*, in Journées de géométrie algébrique d'Orsay, juillet 1992, A. Beauville, O. Debarre, and Y. Laszlo, eds., Astérisque, **218** (1993).

[56] W. Schmidt, *Variation of Hodge structure: the singularities of the period mapping*, Invent. Math., **113** (1981), 45–66.

[57] J. Steenbrink, *Limits of Hodge structures*, Invent. Math., **31** (1976), 229–257.

[58] S. Zucker, *Degenerations of Hodge bundles (after Steenbrink)*, in [**59**], 121–141.

[59] *Topics in Transcendental Algebraic Geometry*, P. Griffiths, ed., Annals of Mathematics Studies 106, Princeton University Press, 1984.

Chapter 4

[60] V. V. Batyrev, *Dual polyhedra and mirror symmetry for Calabi–Yau hypersurfaces*, J. Alg. Geom., **3** (1994), 493–535.

[61] V. V. Batyrev, *Hodge theory of hypersurfaces in toric varieties and recent developments in quantum physics*, Habilitationsschrift, Universität Essen, 1992.

[62] V. V. Batyrev and L. A. Borisov, *Mirror duality and string theoretic Hodge numbers*, to appear in Invent. Math.

[63] V. I. Danilov, *The geometry of toric varieties*, Russ. Math. Surv., **33** (1978), No. 2, 97–154.

[64] V. I. Danilov and A. G. Khovanskii, *Newton polyhedra and an algorithm for computing Hodge–Deligne numbers*, Math. USSR, Izv., **29** (1987), 279–298.

[65] P. Deligne, *Théorie de Hodge II*, Inst. Hautes Études Sci. Publ. Math. (1971), No. 40, 5–57.

[66] W. Fulton, *Introduction to Toric Varieties*, Princeton University Press Study No. 131, 1993.

[67] J. Milnor, *Morse Theory*, Annals of Mathematics Studies No. 51, Princeton University Press, 1963.

[68] M. Reid, *Decomposition of toric morphisms*, in: Arithmetic and Geometry, papers dedicated to I. R. Shafarevich on the occasion of his 60th birthday, Vol. II, Geometry, Progress in Math., **36**, Birkhäuser, 1983, 395–418.

[69] J. H. M. Steenbrink, *Mixed Hodge structure on the vanishing cohomology*, in Real and Complex Singularities, Sijthoff–Noordhoff, Alphen aan den Rijn, 1977, 565–678.

Chapter 5

[70] P. S. Aspinwall and D. Morrison, *Topological field theory and rational curves*, Commun. Math. Phys., **151** (1993), 245–262.

[71] M. Audin, *Cohomologie quantique*, Séminaire Bourbaki, 1995–1996, Astérisque (to appear).

[72] A. Beauville, *Quantum cohomology of complete intersections*, Math. Phys. Analysis and Geom., Vol. 2, 1995.

[73] A. Bertram, *Modular Schubert calculus*, to appear in Advances in Mathematics.

[74] P. Deligne and D. Mumford, *The irreducibility of the space of curves of given genus*, Publ. Math. I.H.E.S., **36** (1969), 75–109.

[75] B. Dubrovin, *Integrable systems in topological field theory*, Nucl. Phys. B, **379** (1992), 627–689.

[76] M. Gromov, *Pseudoholomorphic curves in symplectic manifolds*, Invent. Math., **82** (1985), 307–347.

[77] S. Keel, *Intersection theory of moduli spaces of stable n-pointed rational curves of genus 0*, Trans. Amer. Math. Soc., **330** (1992) 545–574.

[78] M. Kontsevich and Yu. Manin, *Gromov–Witten classes, quantum cohomology, and enumerative geometry*, Commun. Math. Phys., **164** (1994), 525–562.

[79] Yu. Manin, *Generating functions in algebraic geometry and sums over trees*, Proceedings of the conference on "The moduli space of curves," Dijkgraaf, Faber, van der Geer, eds., Progress in Math., **129**, Birkhäuser, 1995.

[80] V. Mathai and D. Quillen, *Superconnections, Thom classes, and equivariant differential forms*, Topology, **25** (1986), No. 1, 85–110.

[81] D. McDuff and D. Salamon, *J-holomorphic curves and quantum cohomology*, University Lecture Series, No. 6, American Mathematical Society, 1994.

[82] Y. Ruan, *Symplectic topology on algebraic threefolds*, J. Diff. Geom., **39** (1994), 215–227.

[83] Y. Ruan and G. Tian, *A mathematical theory of quantum cohomology*, J. Diff. Geom., **42** (1995), 259–367.

[84] C. Vafa, *Topological mirrors and quantum rings*, in [**39**].

[85] C. Voisin, *A mathematical proof of a formula of Aspinwall and Morrison*, Comp. Math., **104** (1996), 135–151.

[86] *Holomorphic curves in symplectic geometry*, M. Audin and J. Lafontaine, eds., Progress in Math., **117**, Birkhäuser, 1994.

Chapter 6

[87] M. F. Atiyah and R. Bott, *The moment map and equivariant cohomology*, Topology, **23** (1984), No. 1, 1–28.

[88] M. Audin, *The topology of torus actions on symplectic manifolds*, Progress in Math., **93**, Birkhäuser, 1991.

[89] A. Floer, *Symplectic fixed points and holomorphic spheres*, Commun. Math. Phys., **120** (1989), 575–611.

[90] A. Floer, *Witten's complex and infinite-dimensional Morse theory*, J. Diff. Geom., **30** (1989), 207–221.

[91] A. B. Givental, *Homological geometry I: projective hypersurfaces*, Selecta Math. (NS), **1** (1995), 325–345.

[92] A. B. Givental, *Homological Geometry and Mirror Symmetry*, Proceedings of the International Congress of Mathematicians, Zürich 1994, Vol. 1, 473–480, Birkhäuser, 1995.

[93] A. Givental and B. Kim, *Quantum cohomology of flag manifolds and Toda lattices*, Commun. Math. Phys., **168** (1995), 609–641.

[94] H. Hofer and D. Salamon, *Floer homology and Novikov rings*, in *A. Floer Memorial Volume*, Progress in Math., **133**, Birkhäuser, 1995.

[95] S. Piunikhin, *Quantum and Floer cohomology have the same ring structure*, to appear in J. Diff. Geom., 1996.

[96] S. Piunikhin, D. Salamon, and M. Schwarz, *Symplectic Floer–Donaldson theory and quantum cohomology*, preprint, University of Warwick, 1995.

[97] C. Viterbo, *The cup-product on the Thom–Smale–Witten complex and Floer cohomology*, in *A. Floer Memorial Volume*, Progress in Math., **133**, Birkhäuser, 1995.

Recent Work

[G] A. Givental, *Equivariant Gromov–Witten invariants*, IMRN, **13** (1996), 613–663.

[BK] S. Barannikov and M. Kontsevich, *Frobenius manifolds and formality of Lie algebras of polyvector fields*, IMRN, 1998, No. 4.

[M] D. Morrison, *Mathematical aspects of mirror symmetry*, Alg-Geom/96 09021.

[SYZ] A. Strominger, S. T. Yau, and E. Zaslow, *Mirror symmetry is T-duality*, Nuclear Physics B **47g** (1996), 243–259.

[BCPP] C. Bini, C. de Corcini, M. Polito, and C. Procesi, *On the work of Givental relative to mirror symmetry*, Appunti della scuola normale superiore di Pisa, 1998.

[BM] K. Behrend and Yu. Manin, *Stacks of stable maps and Gromov–Witten invariants*, Duke Math. J., **85** (1996), No. 1, 1–60.

[BF] K. Behrend and B. Fantecki, *The intrinsic normal cone*, Invent. Math., **128** (1997), 45–88.

[H] N. Hitchin, *The moduli space of complex Lagrangian submanifolds*, Preprint math. DG/9901069.

[Gr] M. Gross, *Special Lagrangian filtrations I: topology*, in: *Integrable Systems and Algebraic Geometry. Proceedings of the Taniguchi Symposium, 1997*, World Scientific, pp. 156–193.

[LR] A. M. Li and Y. Ruan, *Symplectic surgery and Gromov–Witten invariants of Calabi–Yau threefolds I*, to appear in Invent. Math.